GREEN EXTRACTION IN SEPARATION TECHNOLOGY

GREEN EXTRACTION IN SEPARATION TECHNOLOGY

Ali Haghighi Asl
Maryam Khajenoori

Faculty of Chemical, Petroleum and Gas Engineering, Semnan University, Semnan, Iran

CRC Press is an imprint of the
Taylor & Francis Group, an **informa** business

First edition published 2022
by CRC Press
6000 Broken Sound Parkway NW, Suite 300, Boca Raton, FL 33487-2742

and by CRC Press
2 Park Square, Milton Park, Abingdon, Oxon, OX14 4RN

ISBN: 978-1-032-05040-9 (hbk)
ISBN: 978-1-032-05041-6 (pbk)
ISBN: 978-1-003-19577-1 (ebk)

Typeset in Times
by KnowledgeWorks Global Ltd.

CONTENTS

Preface ix

Authors xi

1 New Technologies for Extracting Natural Compounds from Plants 1
- 1.1 Essential Oils 1
- 1.2 Definition of Extraction 3
- 1.3 Definition of Green Extraction 4
- 1.4 Distillation with Water or Steam 4
- 1.5 New Technologies for Extracting Natural Compounds 5
  - 1.5.1 Microwave-Assisted Solvent Extraction (MASE) 5
    - 1.5.1.1 Microwave-Assisted Extraction Systems 6
    - 1.5.1.2 Advantages and Disadvantages of Microwave Extraction Methods 8
    - 1.5.1.3 Researches by Microwave Extraction Method 8
  - 1.5.2 Ultrasound-Assisted Extraction (UAE) 10
    - 1.5.2.1 Ultrasonic Wave Extraction Systems 13
    - 1.5.2.2 Advantages and Disadvantages of Ultrasonic Extraction Method 15
    - 1.5.2.3 Researches by Ultrasonic Extraction Method 16

1.5.3 Instant Controlled Pressure Drop (DIC) Method 17
1.5.3.1 Equipment of DIC Method 18
1.5.3.2 Advantages and Disadvantages of DIC Method 20
1.5.3.3 Research Conducted by DIC Method 20
1.5.4 Supercritical Fluid Extraction 21
1.5.4.1 Supercritical Fluid Extraction Equipment 22
1.5.4.2 Advantages and Disadvantages of SFE 22
1.5.4.3 Research Conducted by SFE Method 24
1.5.5 Subcritical Water Extraction (SWE) 25
1.5.5.1 Subcritical Water Extraction Equipment 25
1.5.5.2 Advantages and Disadvantages of SWE 26
1.5.5.3 Research Done by SWE Method 27
1.6 Conclusion 32
References 32

2 Review of Subcritical Water Extraction (SWE) 37
2.1 Introduction 37
2.2 Subcritical Water Extraction (SWE) 37
2.3 Comparison of SWE and Other Methods 41
2.4 Applications of SWE 44
2.4.1 Extraction of Biologically Active and Nutritional Compounds from Plant and Food Materials 45
2.4.1.1 Antioxidants 45
2.4.1.2 Phenolic Compounds 46
2.4.1.3 Essential Oils 46
2.4.1.4 Other Plant and Food Ingredients 47
2.4.2 Eliminate Organic Contaminants from Food 47
2.5 Economic Study of Subcritical Water Extraction 47
References 48

3 Solubility of Subcritical Water 51
3.1 Introduction 51

3.2 Effective Parameters on Solubility 52
3.2.1 Solvent Type 52
3.2.2 Temperature 53
3.2.3 Flow Rate 53
3.2.4 Pressure 53
3.2.5 Dynamic or Static Mode 53
3.2.6 Concentration of Additives 54
3.3 Solubility Measurement Methods 54
3.3.1 A Review of Laboratory Work Performed 54
3.3.1.1 Static Method 54
3.3.1.2 Dynamic Method 62
3.4 An Overview of the Modeling Performed 66
3.4.1 Experimental and Semi-Experimental Models 67
3.4.2 Dielectric Constant Model 69
3.4.3 Equations of State (EOS) 70
3.4.4 Regular Solution Theory (RST) 71
3.4.5 UNIFAC-Based Models 74
References 78

4 Modeling of Subcritical Water Extraction 85
4.1 Introduction 85
4.2 Investigation of Existing Models in SWE 87
4.2.1 Thermodynamic Model 87
4.2.2 Kinetic Absorption Model 88
4.2.2.1 One-Site Kinetic Desorption Model 88
4.2.2.2 Two-Site Kinetic Desorption Model 89
4.2.3 Thermodynamic Separation with External Mass Transfer Resistance Model 90
4.2.4 Model Based on Differential Mass Balance Equations 93
4.3 Description of the Selected Mathematical Model 97
4.4 Simulation Method 102
4.5 Solution of the Partial Differential Equation 102
4.5.1 Essential Oil Concentration Curve in Bulk Fluid 102
4.5.2 Essential Oil Concentration Curve in Solid 105
4.6 Estimation of Model Parameters and Physical Properties 106

4.6.1 Estimation of the Equilibrium Dissociation Coefficient of the Analyte 107
4.6.2 Estimation of Mass Transfer Coefficient of Bulk Phase 107
4.6.3 Estimation of Solute Diffusion Coefficient in Fluid 108
4.6.4 Physical Properties 109
References 112

Subject Index 115

PREFACE

Research over the past few years has shown that the subcritical water extraction can be used in the case of natural essential oils and foodstuffs, given the growing interest in removing organic solvents from nature. This method is an environment-friendly, inexpensive and non-toxic method of extraction. The equipment required is relatively simple and does not need high pressures like the supercritical fluid extraction method. However, research in this area has been done on a laboratory scale, and still new research needs to be performed. Because of its many benefits over other separation methods, it will be more likely to be produced in the near future for the production of natural essential oils and nutrients.

This book consists of four chapters. In Chapter 1, new strategies for extracting natural compounds from plants are examined. In Chapter 2, research on SWE in the last 15 years is studied. In Chapter 3, a review of the work done in relation to the solubility of different compounds in subcritical water as well as the theoretical content in this field is described in detail. Chapter 4 examines the models used to SWE. Also in this chapter, the principles used in the development of the new model for this process are presented.

There is no specialized book on extracting water with subcritical water as a solvent, and also, we have done a lot of experimental work in this field to make this collection available to researchers and enthusiasts. It is hoped that this collection will be of interest to experts, enthusiasts and scholars, and with their guidance and criticism, will help us to provide the most desirable presentation in future editions.

AUTHORS

Ali Haghighi Asl, PhD, is a professor at Semnan University, Semnan, Iran. He earned a BSc in petroleum engineering at Isfahan University of Technology (IUT), Isfahan, Iran, in 1993, and an MSc in chemical engineering at Tarbiat Modares University, Tehran, Iran, in 1995, and a PhD at Tehran University, Tehran, Iran, in 2001. His research interests include the area of separation processes, extraction processes, subcritical water extraction, and membrane separation.

Maryam Khajenoori, PhD, is an assistant professor at Semnan University, Semnan, Iran. She earned a BSc in polymer engineering at Isfahan University of Technology (IUT), Isfahan, Iran, in 2006, and an MSc and PhD at Semnan University, Semnan, Iran, in 2008 and 2014, respectively. Her current research interests are separation processes, subcritical water extraction, wastewater treatment, membrane processes and production of pharmaceutical nanoparticles.

Chapter 1

NEW TECHNOLOGIES FOR EXTRACTING NATURAL COMPOUNDS FROM PLANTS

1.1 ESSENTIAL OILS

Essential oils or essences are aromatic compounds that are widely used in the perfume, pharmaceutical and food industries. Essential oils are a mixture of more than 200 compounds that can be divided into two general categories. One part is volatile compounds that make up about 90-95% of total oil and contain monoterpene and sesquiterpene hydrocarbons and their oxygen derivatives along with aldehydes, alcohols, aliphatic esters, ketones, lactones and phenols. The other part is non-volatile residue that makes up 5–10% of the oil and contains hydrocarbons, fatty acids, sterols, carotenoids, waxes and flavonoids [1, 2].

Figure 1.1 schematically shows the heterogeneous chemical groups in the essential oils. The terpene component of essential oils has very little effect on the aroma of essential oils, and since terpenes are mainly unsaturated compounds, they decompose by heat, light and oxygen to produce unwanted compounds with an unpleasant odor. The oxygen-containing part of the essential oil is very fragrant, and mainly the main specifications of the essential oil are related to this part. It has oxygen compounds.

Essential oils are classified into three groups: (1) Natural essential oils are products obtained from plant raw materials by one of the extraction methods (distillation, mechanical pressing and extraction with solvent), (2) semi-natural essential oils are products that are formed from a combination of aromatic raw materials and are similar in smell to natural essential oils, (3) artificial essential oils are products that are commercially produced from organic chemicals

Acids	Aldehydes	Alcohols	Ester	Ketones
• Benzoic • Cinnamic • Myristic • Isovaeric	• Citral • Citronellal • Benzaldehyde • Cinnamaldehyde • Vanillin	• Geraniol • Citronellol • Menthol • Linalool • Trponeol • Borneol	• Benzoates • Acetates • Salicylates • Cinnamates	• Camphor • Carvone • Menthone • Pulegone • Thujone
	Essential Oil Constituents			
Oxides	**Phenol Ethers**	**Phenols**	**Hydrocarbons**	**Terpenes**
• Cineol	• Anethol • Safrol	• Eugenol • Thymol • Carvacrol	• Cymene • Myrcene • Sabinene • Storene	• Limonene • Phellandrene • Pinene • Camphene • Cedrene

Figure 1.1 Heterogeneous chemical groups present in essential oil.

similar to natural essential oils and have an odor similar to natural essential oils.

In general, each method of separation from solid matrix consists of two steps: Extraction (such as single-stage solvent extraction, Soxhlet extraction, steam distillation, simultaneous distillation extraction, etc.) and analysis (gas chromatography and gravimetry, etc.). While the analysis step is complete after only 2–20 min, extraction takes at least a few hours. Separation from the solid matrix is often performed by prolonged heating and stirring in the solvent. Therefore, in principle, extraction is the main limiting step, which includes the transfer of desirable compounds in the solvent. The usual solvent extraction method involves 70% of the total separation process time. Therefore, it is very important to shorten this restrictive step. The choice of a method is the result of an agreement between yield, repeatability of the extraction, ease of the procedure, taking into account the cost, time of the process, the degree of automation of the process and safety.

Traditionally, solid matter extraction is performed by the Soxhlet extraction method. This method is obtained by the interaction of the sample with solvent condensed vapors. Soxhlet has been one of the most widely used methods of solid-liquid extraction for a long time and is currently the main reference method. Soxhlet extraction from solids has undeniable advantages such as continuous extraction with repeated infiltration of fresh solvent, no need for treatment stage and the possibility of solvent recovery. However, this method has some disadvantages, including poor extraction of polar compounds, long operating time, large solvent volume, solvent boiling point operation and is unsuitable for unstable thermal analyzes.

These shortcomings have led to the use of new "fast" and "green" techniques in sample preparation, which typically use less solvent and less energy, such as ultrasound-assisted extraction (UAE), supercritical fluid extraction (SFE), microwave extraction, controlled pressure drop process, accelerated solvent extraction (ASE) and subcritical water extraction (SWE).

At present, the extraction and preparation of samples under conditional or non-classical conditions is a dynamic developing area in analytical chemistry. Using these "quick" and "green" sample preparation methods, extraction and distillation can now be performed in a matter of minutes instead of hours with high repeatability, reduced solvent consumption and simple manufacturing, achieving higher purity of the final product, eliminating pre-treatment waste and consuming only a fraction of the energy required against conventional sampling methods, such as Soxhlet extraction, Clevenger, Dean-Stark method.

This chapter provides a brief overview of current knowledge of innovative methods of extracting natural compounds from plants. This study provides the necessary theoretical background and some details about extraction using innovative, fast and green techniques such as ultrasound, microwave, controlled pressure drop process, supercritical fluid extraction, subcritical water extraction, the method of performance, application and their environmental effects.

1.2 DEFINITION OF EXTRACTION

Extraction is one of the most important processes in separation. Extraction of a part of a solid matrix can be considered as a five-step process: Attracting a combination of active sites of the matrix, its diffusion into the matrix; dissolved in extract, diffusion of composition into the extract and collection of extracted components (Figure 1.2). Optimal optimization and control of each stage, especially the extract collection stage, is necessary. In environmental applications (for example, the extraction of pollutants from soils and sediments), the first step is usually to limit the speed, because the effects of the extraction material matrix are very difficult to overcome and predict. For other matrices (for example, plant material), speed may be controlled by solubility or penetration. Therefore, optimization strategies depend on the nature of the matrix.

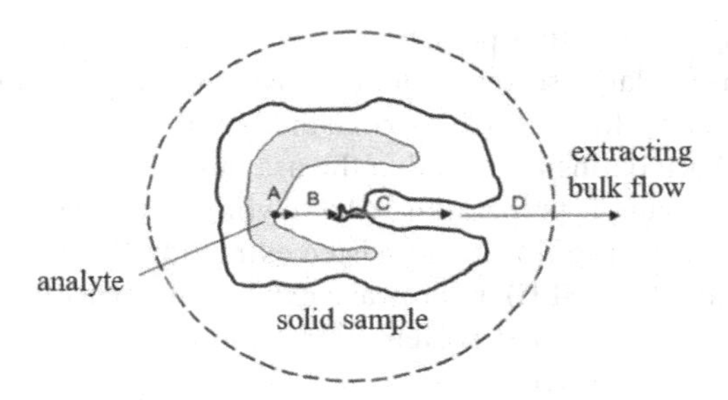

Figure 1.2 Steps of extraction from solid samples.

1.3 DEFINITION OF GREEN EXTRACTION

The general definition of green chemistry is the invention, design and application of chemical products and processes to reduce or eliminate the use and production of hazardous materials. Regarding the green extraction of compounds from plants, this definition has been amended as follows: "Green extraction is based on recognizing and designing extraction processes that reduce energy consumption, increase the use of alternative solvents, and ensure the safety of a high-quality product or extract" [3].

1.4 DISTILLATION WITH WATER OR STEAM

One of the most common methods of extracting essential oils is distillation with water or steam. Based on the water or steam distillation method, three systems of water distillation, water-steam distillation and steam distillation have been designed and built so far such that steam distillation system is generally used on an industrial scale. In this way, water vapor is placed at low pressure (1.0 bar) and the steam, after passing through the plant mass and extracting the compounds, collects its essential oil in the cooling part of the system.

Disadvantages of this method are the risk of loss of volatile compounds against heat, the impracticality of the process automatically and the long time for extraction. However, in order to prevent the loss of volatile compounds against heat, vacuum distillation is used, which is the process of distillation of water vapor that is performed at low pressure. However, this method still has two significant disadvantages mentioned above.

1.5 NEW TECHNOLOGIES FOR EXTRACTING NATURAL COMPOUNDS

As mentioned, traditional methods have given way to new methods due to their time-consuming extraction and high solvent consumption. Therefore, there is a great demand for new extraction methods with shorter time, lower solvent consumption and environmental protection. New methods of extracting essential oils such as extraction with ultrasound, extraction with microwave, extraction with supercritical fluid and extraction with subcritical water are very fast and effective for extracting plant compounds.

1.5.1 Microwave-Assisted Solvent Extraction (MASE)

The use of microwave energy was first reported in 1986 simultaneously by Gedbye in organic synthesis [4] and by Ganzler for the extraction of biological samples and analysis of organic compounds [5]. Since then, several laboratories have studied the combined and analytical capabilities of the microwave as a non-classical source of energy. Several categories of compounds such as essential oils, fragrances, pigments, antioxidants and other organic compounds have been effectively extracted from a variety of matrices (mainly animal, food and plant tissues). Advances in microwave extraction have provided two categories of solutions: Microwave-assisted solvent extraction (MASE) and solvent-free microwave extraction (SFME).

MASE was first used to extract several compounds (citrus, aromatic plants, cereals, etc.). Many categories of compounds such as perfumes, antioxidants, dyes, bio phenols and other primary and secondary metabolites have been effectively extracted from a variety of matrices. The technique was invented in 1990 as a microwave-assisted process. Microwaves are electromagnetic radiation with a frequency of 0.3–300 GHz.

Extraction with the help of microwaves is based on the absorption of microwave energy by polar molecules of chemical compounds. The absorbed energy is proportional to the dielectric constant of the object and causes a bipolar rotation in the electric field (usually 2.45 GHz). Extraction takes place at temperatures between 150 and 190°C. Hot solvent allows for rapid extraction of stable thermal analyzes. The efficiency of microwave extraction depends on the properties and volume of the solvent, the ratio of the solvent volume to the extraction material, the power of the microwave and the duration of irradiation, the size of the plant particles and the state of the plant material and especially the relative dielectric constant. For compressed microwave-assisted extraction (MAE), a clear microwave container and solvent with high dielectric constant is used, which strongly absorbs microwaves. At higher dielectric constant, more energy is absorbed by the molecules and a faster solvent reaches the boiling point. Therefore, increasing the temperature and pressure makes it easier to extract compounds from plants. Solutions with low dielectric constant can also be used at atmospheric pressure. In this case, extraction takes place in open containers, which is why the solvent absorbs a small amount of energy. The sample matrix is heated and analyzes are transferred to a cooling solvent. This method is used to extract unstable low-polar thermal analyzes.

Plants are usually immersed in an insoluble microwave solvent, such as hexane, and irradiated with microwave energy. Although in most cases dry plant material is used for extraction, plant cells still have small amounts of moisture that are used as a target for heating microwaves. Humidity is also exposed to intense microwave heating when exposed to extreme heat stress and high local pressure; inside the plant cells, they heat up, evaporate and produce a lot of pressure due to the swelling of the plant cell on the cell wall. Pressure presses on the cell wall from the inside and eventually causes them to rupture more quickly than conventional extraction, which facilitates the leakage of active components from the damaged cells to the surrounding solvent. And so it improves the efficiency of extracting plant components. High temperatures generated by radiation from microwaves can hydrolyze any cellulose bond that is a major component of the cell wall, and instead convert it into soluble components in a matter of minutes. High temperatures in cell walls during extraction by microwaves increase cellulose dehydration and reduce its mechanical strength, which facilitates easy solvent access to intracellular compounds.

1.5.1.1 Microwave-Assisted Extraction Systems

Solvent-free microwave hydro-distillation (SFME) is a new method of extraction that was invented in 2004 to obtain essential oils from plants (Figure 1.3A). The SFME is the main component of microwave heating and atmospheric pressure distillation. Based on a relatively simple principle, this method involves placing plant material in a microwave reactor without adding any solvents. The internal heat of

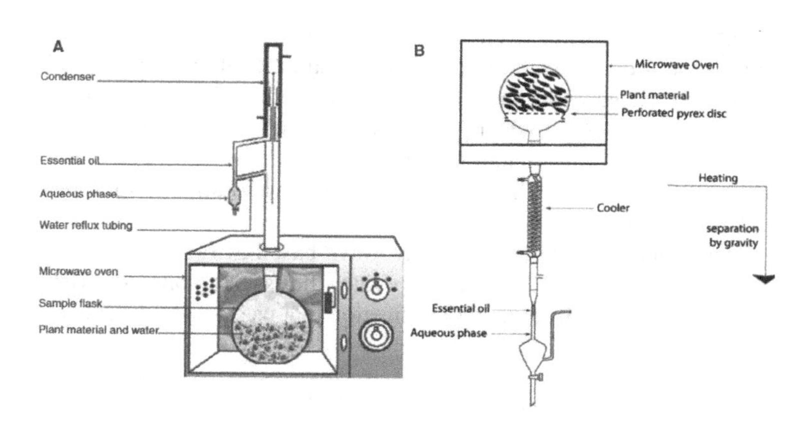

Figure 1.3 (A) Hydro-distillation by microwaves. (B) Microwave hydro-diffusion and gravity.

the water in the plant material causes the swelling of the plant cells and the rupture of the glands and aromatic vessels. Therefore, the process of releasing essential oil is done by evaporating the water in the plant material. A cooling system outside the microwave oven continuously compresses the extract. In order to return the available water to the plant material, the excess water is returned to the extraction cell. The higher frequency of oxygenated compounds in essential oils is due to the rapid heating of polar materials with microwaves, and only available water is used to prevent the decomposition of the main oxygenated compounds by thermal and hydraulic reactions.

The new and green technique of microwave hydro-diffusion and gravity (MHG) for extracting essential oils was invented in 2008 (Figure 1.3B). This method of green extraction is a combination of main reverse microwave distillation, microwave heating and earth gravity at atmospheric pressure. This method also involves placing the plant material in a microwave reactor without adding any solvents or water. The internal heat of the water in the plant material causes the swelling of the plant cells and the rupture of the glands and aromatic vessels. As a result, the microwave heating process releases the essential oil and water from inside to outside the plant material. This physical phenomenon, known as hydro-diffusion, allows the extract (water and essential oil) to be dispersed outside the plant material, gravitationally extracted from the outside of the microwave reactor and dropped through the perforated piercing disc. A cooling system outside the microwave oven constantly cools the extract. The water and essential oil collected in a container called a "fluorescent flask" are usually separated. The essential oil is lighter than water, so it floats on top while the water goes down and they can be easily separated. It is important to note that this green method allows essential oils to be extracted without distillation and evaporation, which are common methods.

The MHG method is neither microwave-based extraction that uses organic solvents nor SFME that evaporates the essential oil with existing water, nor is hydrophobic modification that uses large amounts of water and energy.

1.5.1.2 Advantages and Disadvantages of Microwave Extraction Methods

Some of the main advantages of microwave extraction for product extraction from plant materials compared to traditional methods include shorter extraction time, reduction of extraction device size, faster energy transfer, reduction of thermal gradient, easy control of sample heating, reduction of amount consumer solvent, improving extraction efficiency, performing the process automatically, creating turbulence during extraction, simplicity of the process and low cost. One of the disadvantages of this method is that the efficiency of microwaves is very low and weak when the target compounds and solvent are non-polar or when they are volatile. Therefore, when used to extract non-polar analytes from non-polar materials, it is necessary to use solvents with bipolar moment greater than zero (n-hexane or isooctane can be replaced with dichloromethane or a mixture of acetone and n-hexane). Also, the use of high microwave energies may lead to isomerization or degradation of the compound, and a filter or centrifugal force is required to remove solid residues during the process.

1.5.1.3 Researches by Microwave Extraction Method

The feasibility of the microwave process in the preparation of samples from different matrices has been investigated, some of which are mentioned in this section. Lucchesi et al. (2004) extracted the essential oils of three aromatic plants, basil, peppermint and thyme with SFME [6]. Using this method, the separation and concentration of volatile compounds were performed in one step, without adding any solvent or water. The extracted essential oils were rich in the amount of oxygenated compounds of eugenol (43.2%) in basil, carvone (64.9%) in mint and thymol (51%) in thyme, so that they were comparable to the conventional method. In fact, the higher frequency of oxygenated compounds in the essential oil is due to the rapid heating of polar materials by microwave and the use of less water, which prevents the decomposition of the main oxygenated compounds by thermal and hydraulic reactions.

Hemwimon et al. (2007) made a comparison between microwave extraction of antioxidants such as anthraquinones from *Morinda citrifolia* root and extraction by other conventional methods (soaking and Soxhlet) and ultrasonic extraction [7]. Extraction efficiency using MAE (720 W, 15 min, 60°C, 10 mL ethanol water (80:20)) higher than the soaking values (3 days, 25°C, 10 mL ethanol) and ultrasound (15.7 W, 60 min, 60°C, 10 mL ethanol) was also comparable to Soxhlet (4 h, boiling point, 200 mL ethanol). The main reason for recovering higher anthraquinone with MAE was bipolar rotation in polar solvents in the microwave range. The extract obtained from MAE had higher antioxidant activity than ultrasound and soaking. Exposure of the soaking extract to adverse conditions such as less light and oxygen resulted in lower activity due to its longer extraction time. Ultrasound, on the other hand, did not require a long extraction time, but due to the formation of free radicals by ultrasound, it caused oxidation and destruction of anthraquinones.

Soxhlet extraction using concentrated microwaves by Pérez-Serradilla et al. (2007) was used to determine oak oil (Trans fatty acid-free oil) and the extraction was performed in 30 min, which was much less than the time required by the Soxhlet reference methods (8 h) and stirrer (56 h) [8]. One of the common drawbacks of all these methods, in addition to environmental pollution, is the use of n-hexane solvent. However, this problem has also been solved with the development of a new method: The extraction of Soxhlet combined with a microwave using di-limonene as a bio-solvent derived from the skin of a lemon, which is a new alternative to organic solvents. This method, along with reducing the environmental pressure, offers the best performance in less time with more solvent recovery.

Lianfu and Zelong (2008) explained the extraction of Lycopene from tomatoes by combining two innovative techniques of extraction by microwave and ultrasound compared to extraction by ultrasound alone [9]. The results obtained with ultrasound did not show an increase in Lycopene efficiency (89.4%) in a shorter time. The reason for this was the production of hydroxyl radicals by sonic cavitation in the extract due to the presence of a small amount of water in the extract and the resulting decomposition of Lycopene. However, when the extraction was performed by microwave and ultrasound, the efficiency (97.4%) increased with shorter time and less solvent consumption due to acoustic cavitation and rapid microwave heating.

Bousbia et al. (2009) compared the extraction efficiency of essential oil from rosemary leaves by two methods of traditional water distillation and hydro-diffusion by microwave and gravity [10]. Distillation extraction method using microwaves and gravity had shorter extraction time and more valuable essential oil (with a large amount of oxygenated compounds) than the water distillation extraction method. Gujara et al. (2010) used the microwave extraction method to extract thymol from *Trachyspermum ammi* seeds [11]. The maximum thymol extraction time was 45 min, the optimum MAE conditions were 45°C and the ratio of sample weight to solvent volume was 1:30.

1.5.2 Ultrasound-Assisted Extraction (UAE)

Ultrasound is a mechanical wave that requires an elastic medium and sounds with different wave frequencies to scatter (Figure 1.4). Sounds in the human frequency range from 16 Hz to 16–20 kHz, while ultrasound has frequencies higher than human hearing but lower than microwave frequencies (from 20 kHz to 10 MHz) [12]. Ultrasound can be divided into two groups according to frequency: Diagnostic and power ultrasound. Diagnostic ultrasound (low power and high frequency) is in the range of 5–10 MHz and is used in various fields such as medical imaging or even to diagnose defects (plastic graft inspection). High-power ultrasound (low-frequency ultrasound) is used to produce physical and chemical effects in the environment. They are used in sonochemistry (to facilitate or accelerate chemical reactions), agriculture (water dispersion) or in industry (plastic cutting and welding).

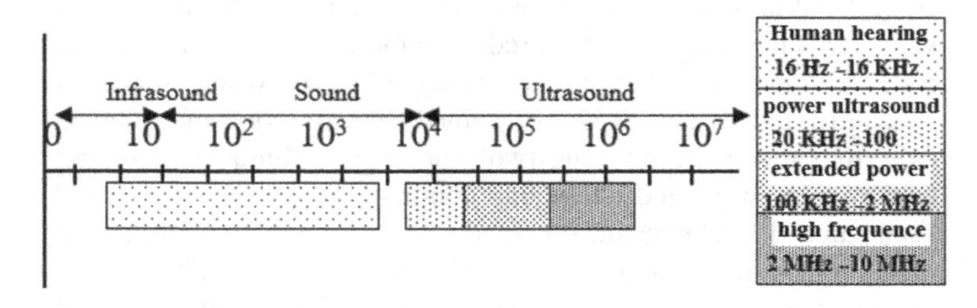

Figure 1.4 Frequency ranges.

The effects of ultrasound in the extraction medium are attributed to the phenomenon of cavitation. As a sound wave passes through an elastic medium (Figure 1.5), the longitudinal displacement of the particles acts as a piston on the surface of the medium, and as a result, a sequence of compression and rarefaction steps takes place [12]. At the molecular scale, molecules are temporarily detached from their original position and move like a sound wave, colliding with surrounding molecules. In the evolution phase, the first group of molecules are pulled back and forth to their original position and kinetic energy helps them move. These rarefaction zones are created in the environment, and since each medium has a critical molecular distance, when this distance is too large, the molecular interactions are broken and holes are created in the liquid. The cavities created in the environment are cavitation bubbles caused by ultrasound. In fact, these cavitation bubbles are able to grow during the dilution process and decrease in size during compression cycles.

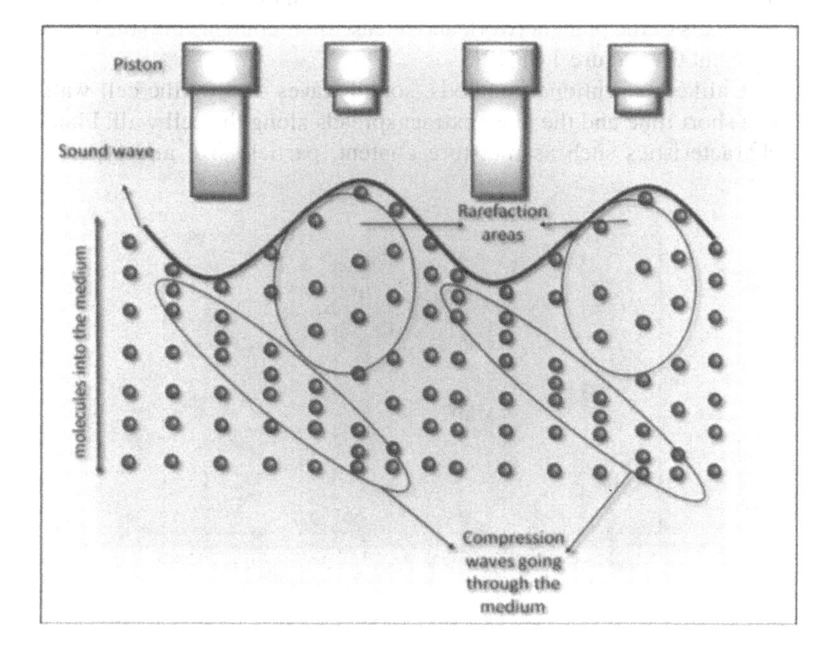

Figure 1.5 Compact and open cycles created by sound waves.

(Source: Pingret, D.; Fabiano-Tixier, A.S. and Chemat, F. (2012). Academic Press, Oxford, 441–455).

When these bubbles reach a critical point, they disintegrate during the contraction cycle, releasing large amounts of energy. Temperature and pressure at the moment of disintegration are estimated at up to 5000 K and 5000 atm in an ultrasonic bath at room temperature. Creating these hotspots can dramatically accelerate chemical reactions in the environment. When these bubbles disintegrate on the surface of the solid, the high pressure and temperature released directly produces microjets and shock waves on the solid surface. The impact of these microjets on the surface causes abrasion, breakage and destruction.

Ultrasound facilitates and accelerates the process of extraction of plant compounds, i.e., swelling of the tissue in order to absorb the solvent and also the exit of the compounds from the tissue to the solvent by creating pores in the cell wall and improving the diffusion and mass transfer. The cavitation bubble produced near the surface of the plant material (a) disintegrates during the contraction cycle (b) and introduces the microjet directly to the surface (b and c). The high pressure and temperature involved in this process will tear the cell walls of the plant network and release their contents into the environment (d) (Figure 1.6) [12].

Unlike conventional methods, sound waves destroy the cell wall in a short time and the plant extract spreads along the cell wall. Plant characteristics such as moisture content, particle size and type of

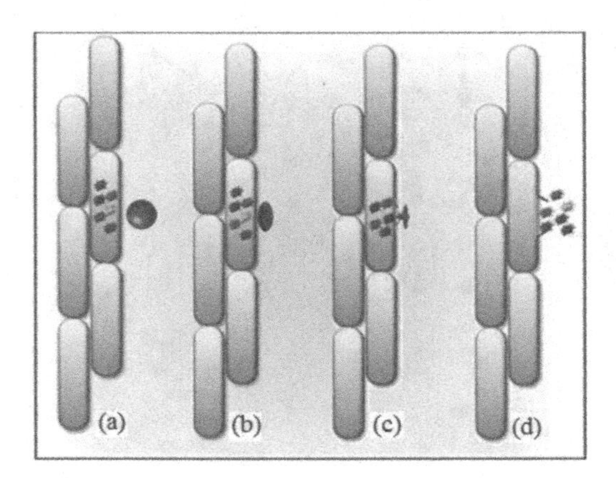

Figure 1.6 Cavitation bubble collapse and release of plant material.

(Source: Pingret, D.; Fabiano-Tixier, A.S. and Chemat, F. (2012). Academic Press, Oxford, 441–455).

solvent used are important in order to obtain efficient and effective extraction. In addition, many factors, including frequency, pressure, temperature and time, affect the performance of sound waves.

1.5.2.1 Ultrasonic Wave Extraction Systems

The most common ultrasonic equipment for extraction purposes from plant sources is ultrasonic purification bath and probe systems that can be implemented on an industrial and laboratory scale. For small extraction volumes, ultrasound vibration with a tip immersed in the liquid may be sufficient. Most of the fluid is in the ultrasound bath or in the recycled-flow sonoreactor (Figure 1.7) [12].

Binding of the UAE to the analysis steps, which can be overcome by pre-analysis dilution, has not yet been performed despite its ease of implementation. In this method, the extract must be driven into a continuous manifold to access the analysis process online, which includes preconcentration, derivatization, purification and detection

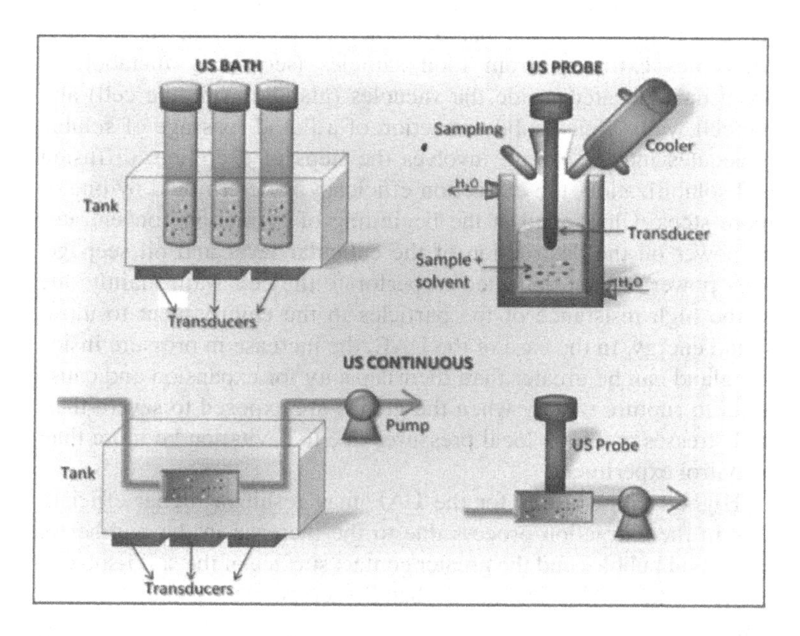

Figure 1.7 Types of equipment for extraction systems using ultrasound.

(Source: Pingret, D.; Fabiano-Tixier, A.S. and Chemat, F. (2012). Academic Press, Oxford, 441–455).

(using flame atomic-absorption spectrometer (FAAS), gas chromatography-mass spectrometry (GC-MS) or other methods).

While much of the research effort in the UAE has focused on ultrasound itself, some studies have examined the connection between ultrasound and other methods. For example, UAE has been used in combination with microwave energy, supercritical fluid extraction or simply by conventional methods such as Soxhlet extraction. When combined with supercritical fluid extraction, UAE increases the mass transfer of beneficial species from the solid phase to the solvent used for extraction. Also, the extraction of Soxhlet can be modified with ultrasound and a siphon is applied in the area of the cartridge before the route to allow the removal of fat parts from very compact matrix. The combined efficiency of microwave and ultrasound has been clearly demonstrated in applications such as copper extraction and the Kjeldahl method for determining total nitrogen in food.

Important physical parameters related to the UAE include ultrasound power, temperature and extraction time, which affect not only the extraction efficiency but also the composition of the extract. Molecules extracted from food samples (secondary metabolites) are usually located inside the vacuoles (glands inside the cell) and the cell walls. Since solid extraction of a liquid (passage of soluble molecules into a solvent) involves the steps of excretion, diffusion and solubilization, the extraction efficiency is determined by one or more steps. Ultrasound at the beginning of extraction concentrates its power on the destruction of the cuticular layer and oil seepage. This power is then deflected to perforate the cell wall, mainly due to the high resistance of the particles in the environment to ultrasound energy. In the case of the UAE, the increase in pressure inside the gland can be greater than their capacity for expansion and cause them to rupture rapidly when the glands are exposed to severe thermal stresses and high local pressures due to cavitation be more than a control experiment.

Higher temperatures for the UAE may result in higher efficiencies in the extraction process due to the increase in the number of cavitation bubbles and the greater contact surface of the solid-solvent. However, this effect decreases when the temperature is near the boiling point of the solvent. Temperature is also important to prevent the degradation of thermally unstable compounds. Ultrasound power is one of the parameters for optimization in order to reach an agreement between extraction time and solvent volume. In general, the highest UAE efficiency, in terms of yield and extract composition, can be achieved by increasing the strength of the ultrasound, reducing the moisture of the food matrix to increase the solid-solvent contact, and optimizing the temperature for a shorter extraction time. The most common frequencies used in sonochemistry are between 20 and 40 KHz. At higher frequencies, cavitation will be more difficult to excite, so cavitation bubbles will need a small delay during the start of the dilution cycle. The higher the frequency, the shorter the evolution cycle lengths, thus reducing the likelihood of bubbles forming.

Cavitation bubbles contain not only a vacuum but also some of the liquid vapor in which they are formed. If there is no gas in the liquid, the bubbles will be significantly more difficult to form. Solvent-dissolved gases act as the nucleus of a new cavitation bubble, thus increasing the rate of cavitation bubble formation. On the other hand, as cavitation bubbles are facilitated, they grow faster and the boiling point of the solvent may change. If the bubbles grow too fast, they will not have time to disintegrate and the liquid will boil without cavitation. The easy production of bubbles in a liquid containing a large amount of dissolved gases indicates why ultrasonic baths are used to degas the liquid.

1.5.2.2 Advantages and Disadvantages of Ultrasonic Extraction Method

The advantages of ultrasonic extraction are the increase of system polarity (including extractor, analytes and matrix) and the increase of extraction efficiency with cavitation, which can be similar or larger than Soxhlet extraction. Ultrasonic extraction allows the addition of an auxiliary extractor and increases the polarity of the liquid phase. Ultrasound can reduce the operating temperature and allow the extraction of heat-sensitive compounds that change under the operating conditions of the Soxhlet extraction. Extraction time is shorter than Soxhlet extraction. One of the advantages of ultrasonic extraction over microwave extraction is that it is faster and easier so less operations are involved, and as a result it is less polluted. In acid absorption, the ultrasonic method is safer than the microwave, so it does not require high temperatures and pressures. Ultrasonic extraction requires much simpler equipment than supercritical fluid extraction. Therefore, the cost of the whole extraction process is much lower. Ultrasonic extraction can be used with any solvent to extract a wide range of natural compounds. On the other hand, supercritical fluid extraction uses CO_2 exclusively for extraction, so its range is limited to non-polar analytes.

One of the disadvantages of ultrasonic extraction compared to Soxhlet method is the lack of solvent renewability in discontinuous systems during the process. Therefore, its efficiency is a function of the distribution coefficient. On the other hand, rinsing and refining after extraction is longer than the total process time and increases the solvent consumption, and there is a possibility of loss or contamination of the extract during handling. Soxhlet extraction is more reproducible. Ultrasonic extraction is usually weaker than microwave extraction, so that the surface of the ultrasonic probe surface can alter the extraction efficiency. Particle size is an important factor in ultrasound applications. The supercritical fluid extraction method is simpler and faster than some liquid solvent ultrasound methods. Unlike some solvents used for ultrasound (such as cyclohexane, tetrahydrofuran and binary mixtures such as dichloromethane and acetone), supercritical CO_2 is not hazardous to the environment. SFE methods are usually more accurate than ultrasonic extraction, possibly due to their use of the bath instead of the probe-type system.

1.5.2.3 Researches by Ultrasonic Extraction Method

The application of ultrasound as a laboratory method for the extraction of plant materials has been extensively studied. The range of extraction applications published by this method includes plant active ingredients, oils, proteins and bioactive compounds from plant materials [13]. Anthraquinones are active compounds that have therapeutic effects and are used in anti-cancer drugs. Hemwimol et al. (2006) investigated the use of ultrasonic extraction to improve the extraction of anthraquinones from *Morinda citrifolia* root with solvent [14]. Ultrasonic extraction in a system (methanol-water) resulted in a 75% reduction in extraction time and efficiency compared to samples not treated with ultrasound.

Martino et al. (2006) conducted research on clover and investigated the effect of microwave, ultrasound and Soxhlet methods on the extraction of coumarin and similar compounds [15]. The best results were obtained for the microwave method (with 50% aqueous ethanol, two 5-min heat cycles and a temperature of 50°C with a closed chamber microwave system). Also, in the study of different times (10–180 min) and different solvents (50% ethanol, 50% methanol and boiling water) on the extraction of these compounds using ultrasound bath, the best case for a time of 60 min and with the solvent of aqueous ethanol was 50%, which had a higher extraction efficiency compared to Soxhlet method. In general, studies show that in the presence of high-power ultrasound, the cell wall and plant tissues are destroyed and more antioxidant compounds (polyphenols and tocopherols) and pigments (chlorophyll and carotenoids) get into the oil, which increases their nutritional value.

Jacques et al. (2007) investigated the chemical composition of tea extract extracts (leaves of *Ilex paraguariensis*) by ultrasonic extraction [16]. The effect of using ultrasound waves improved the efficiency of caffeine and palmitic acid in methanol solvent. Also, in a study conducted by Wang et al. (2008) on the optimization of extraction conditions of phenolic compounds from wheat husk using ultrasonic bath, the best extraction conditions were ethanol concentration of 64%, temperature of 60°C and time 25 min, extraction time was the most important parameter for the process [17].

In 2012, Kongkiatpaiboon and Gritsanapan studied the extraction of the pesticide dihydrostemofoline alkaloids from *Stemona Collinsai* root extract by five different extraction methods (ultrasound, reflux, Soxhlet, soaking and infiltration) with 70% ethanol [18]. Ultrasound and reflux were shown to have the highest efficiencies in the extraction of dihydrostemofoline. Increasing heat or ultrasonic energy during the extraction process can reduce extraction time and help increase efficiency.

1.5.3 Instant Controlled Pressure Drop (DIC) Method

The instantaneous controlled pressure drop process was developed by Allaf et al. in 1988. DIC extraction is based on fundamental studies of thermodynamics [19]. This method consists of a thermomechanical process that results from the product being exposed to the rapid transfer of high vapor pressure to the vacuum. DIC extraction usually begins with the creation of vacuum conditions, which is done by injecting steam into the material and holding it in such a state for a few seconds and then proceeding to a sudden drop in pressure to vacuum (about 5 kPa at a rate higher than 0.5 MPa/s). With a sudden decrease in pressure, rapid automatic evaporation of moisture will occur inside the material, causing swelling and tissue change, leading to greater porosity, an increase in specific surface area and a decrease in penetration resistance. With short extraction time (a few seconds) and instantaneous decrease in temperature after pressure drop, further thermal decomposition is prevented and the final extract is obtained with excellent quality.

1.5.3.1 Equipment of DIC Method

The DIC reactor consists of four main elements, namely (1) an autoclave with a thermal jacket, also called a process extraction vessel, in which samples are placed and extracted; (2) a vacuum system, which mainly consists of a vacuum tank with a volume 50 times larger than the process container; (3) suitable vacuum pump and pressure relief system with pneumatic spherical valves, the autoclave is separated from the vacuum tank by which the extract is cooled by a dense double-walled jacket and recovered through a trap; (4) water rotation pump that maintains the tank pressure of about 5 kPa. Figure 1.8 shows a schematic diagram of the DIC device [12]. Initially, the wet

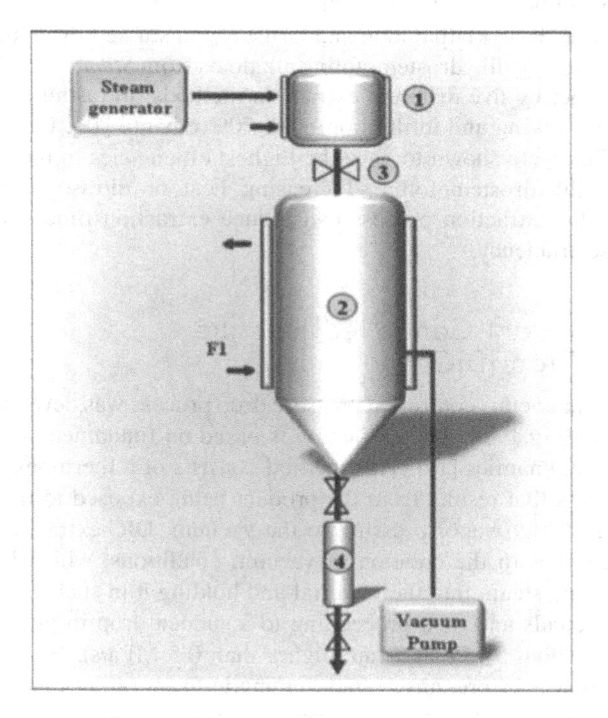

Figure 1.8 Schematic of DIC reactor. (1) Autoclave with thermal jacket; (2) vacuum tank with cooling water jacket; (3) instant controlled pressure drop valve; (4) extract container, (F1) cooling water flow.

(Source: Pingret, D.; Fabiano-Tixier, A.S. and Chemat, F. (2012). Academic Press, Oxford, 441–455).

product is placed in an autoclave at atmospheric pressure before vacuum adjustment. The connection between the autoclave and the vacuum tank is established by the rapid opening of the spherical valve. The initial vacuum facilitates the penetration of the thermal fluid and increases the heat transfer to the product. After closing this valve, the autoclave is filled with steam until the process pressure. After a certain time at a constant process pressure, the pneumatic valve opens immediately (in less than 0.2 seconds), resulting in a sudden drop in pressure inside the autoclave. At this point, the temperature in the autoclave depends on the saturated vapor temperature at the autoclave operating pressure. After the vacuum period, the spherical valve closes and atmospheric pressure is obtained. The concentrated vapor mixture and the extracted metabolites are recovered in a special container placed under the vacuum tank.

The pressure characteristics of the DIC method are shown in Figure 1.9 [12]. In the multi-cycle DIC process, steam is injected again after step (f) and the pressure is controlled for time (d). This means that n cycles contain n repetitions of steps (c) to (f). The last cycle ends with step (g). The total heating time is equal to the heating time of all cycles (n × d). The heating time of each cycle (d) is measured after the constant pressure level is reached. Due to the short steps (c), (e) and (f), the whole process time is a bit longer.

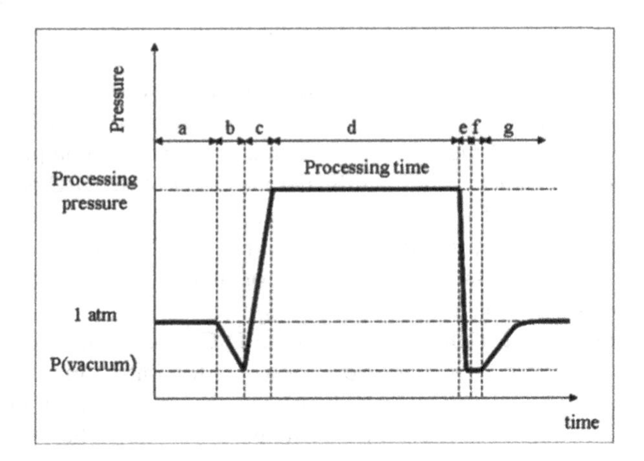

Figure 1.9 DIC process cycle time pressure profile.

(Source: Pingret, D.; Fabiano-Tixier, A.S. and Chemat, F. (2012). Academic Press, Oxford, 441–455).

1.5.3.2 Advantages and Disadvantages of DIC Method

One of its advantages over other extraction processes is the short contact time of the essential oil with the hot areas of the device, which prevents harmful thermal reactions of various molecules. Sudden pressure changes also allow for a rapid release of the essential oil due to rupture of the oil-containing glands. DIC process in terms of speed (a few minutes vs. a few hours), the efficiency of the essential oil (comparable or even higher) is more efficient and contains higher oxygenated compounds. In addition, this method requires only a very small amount of solvent (50–200 mL). The most important drawback of this method is the decomposition of thermally unstable compounds.

1.5.3.3 Research Conducted by DIC Method

The feasibility of DIC process in extraction from different networks has been investigated, some of which are mentioned in this section. Kristiawan et al. (2008) used the DIC process for dried flowers of *Indonesian Cananga* with the aim of separating the essential oil [20]. The effect of number of DIC cycles (1–9) and heating time (4.3–15.7 min) on access to oily compounds at three levels of vapor pressure (0.28, 0.4 and 0.6 MPa) was investigated. The amount of volatiles in the essential oil increased with increasing pressure and the number of cycles. Fast DIC process (0.6 MPa, 8 cycles, 6 min) had better results than steam distillation (16 h) in essential oil yield (2.8% vs. 2.5%) and oxygenated content (72.5% vs. 61.7%).

Ben Amor and Allaf have studied the extraction of anthocyanins from Roselle calyces using DIC [21]. This study showed that DIC process increases the kinetics and extraction efficiency of compounds. The same researchers also investigated the effect of the DIC process on the extraction of oligosaccharides (such as acacia and ciceritol) from the seeds of the *Tephrosia purpurea* plant. The DIC process is an efficient extraction method in terms of time (1 h extraction time instead of 4 h for conventional processes).

Besombes et al. (2010) performed the DIC process in laboratory devices as well as a pilot sample to prove its feasibility, identify the optimal process conditions and determine the energy consumption and the amount of water used for such operations [22]. The obtained lavandin essential oil was studied and modeled using DIC extraction process and its quantity and quality were compared by conventional hydro-distillation method. The most important difference between these two essential oils was the efficiency of 4.25 vs. 2.3 g of essential oil per 100 g of raw material and at the time of extraction 480 seconds versus a few hours for DIC and hydro-distillation operations, respectively.

Pedrosa et al. (2012) used the DIC process to investigate the effect of this process on the content of nutritionally active factors (NAFs) in soybeans, lentils, peas and roasted peanuts [23]. Unprocessed (control) and processed samples (by DIC process under different pressures and time conditions) were analyzed for oligosaccharides, inositol phosphates, Trypsin inhibitors and lectins. The effect of DIC treatment on NAFs in cereal grains showed that this process significantly reduces many of these components. The optimal conditions for the DIC process in grains were a pressure of 6 bar and a time of 1 min. The main advantages of DIC are short processing time and application of whole grains for industrial applications.

1.5.4 Supercritical Fluid Extraction

A supercritical state is achieved when the temperature and pressure of a material exceed the value of its critical temperature and pressure. The properties of supercritical fluid are the same as those of gases and liquids. The advantages of supercritical fluids over liquid solvents are: (1) The dissolution strength of the supercritical fluid depends on its density, which is adjusted by changing the pressure and temperature, and (2) the supercritical fluid has a higher penetration coefficient, lower viscosity and lower surface tension than the liquid solvent. This leads to a more desirable mass transfer.

The choice of solvent is very important for the development of the supercritical fluid extraction process. So far, various materials have been used as solvents in the supercritical state, including hydrocarbons such as hexane, pentane, butane, nitrous oxide, sulfur hexafluoride, many fluorocarbons, water and some alcohols, such as methanol and ethanol. However, carbon dioxide (CO_2) has become very common as a supercritical fluid due to its properties such as non-toxicity, cheapness and availability, and finally suitable critical conditions for this material (pressure 70 bar and temperature close to ambient temperature).

Many studies with supercritical fluids have been performed at pressures of less than 350 bar. CO_2 dissolves many organic compounds under supercritical conditions. Compounds such as unsaturated fatty acids, terpenes and many other substances in plant essential oils can be extracted with supercritical CO_2 under these conditions. The critical pressure for CO_2 is 72.9 bar and the critical temperature is about 31.2°C. With increasing pressure and temperature, the solubility and consequently the extraction capacity of supercritical fluid increases and thus the extraction efficiency increases. However, it should be noted that in addition to high efficiency, the composition of the extracted material is also important and to achieve the required percentage composition in the extracted essential oil, the extraction conditions must be optimized. Resistance to mass transfer due to the structures of the raw material as well as the specific conditions of the materials used is another factor that plays an important role in the extraction process. In general, in order to achieve the best requirements for the product, a proper understanding of the thermodynamics (solubility) and kinetics (mass transfer rate) of the solid and the supercritical phase must be established.

1.5.4.1 Supercritical Fluid Extraction Equipment

In terms of shape, supercritical fluid extraction devices can be divided into several groups. Many of these devices consist of one, two or more extractors (Figure 1.10) [24]. The use of several separators makes it possible to divide the extracted material into several parts (in terms of percentage composition) and by properly adjusting the pressure and temperature of the separators, various products with desired components are obtained. For example, if CO_2 is at a temperature of 50°C and a pressure of 90 bar, compounds that are highly soluble, such as unsaturated oils, will also dissolve in it. Now, if CO_2 is at 300 bar and 50°C, substances that have less solubility, such as antioxidants, are also dissolved in it, and in this way, these two substances can be extracted separately. Another important point that needs to be mentioned here is the role of co-solvent. Co-solvent increases the solubility of supercritical fluid for polar materials and thus at lower pressures the desired materials can be extracted.

1.5.4.2 Advantages and Disadvantages of SFE

Extraction with supercritical fluid of CO_2 has significant advantages over others. It has the usual extraction methods. For example, (1) extraction with supercritical fluid due to the solubility of chemical material in supercritical fluid can be controlled by changing the pressure or temperature of the fluid and provide the possibility of continuous modification of the selectivity of components; (2) in addition, because the solvent strength of the supercritical fluid is proportional to its density, a selective extraction can be obtained using a pressure or temperature gradient, and finally the solvent-free material is

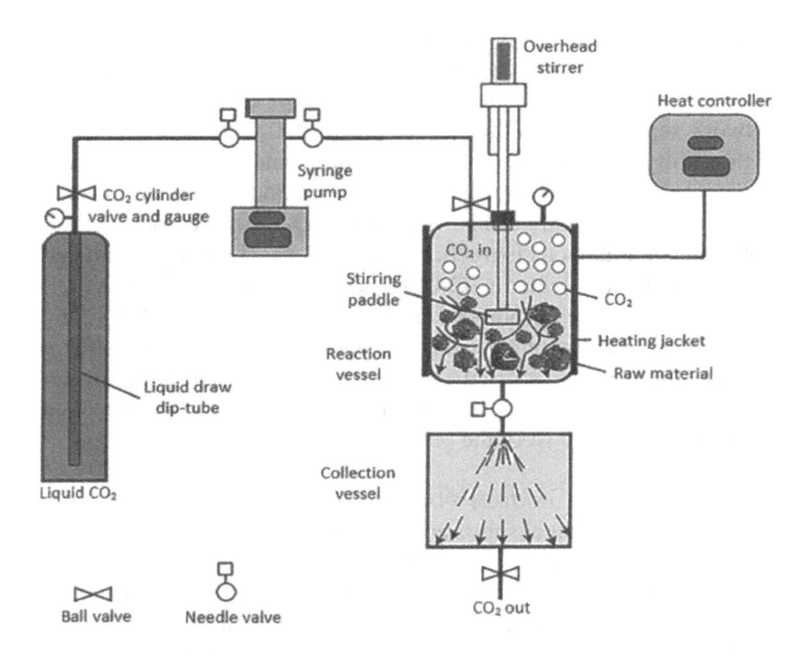

Figure 1.10 Diagram of a laboratory unit extracted with supercritical fluid equipped with two separating cells.

(Source: Khaw, K.Y. et al. (2017). *Molecules* 22: 1186–1208).

extracted after the decompression step is obtained; (3) the penetration coefficient of the supercritical fluid is one or two times higher than the liquid state, which allows high mass transfer, which results in a higher extraction rate than conventional solvent extractions; (4) extraction with supercritical fluid of CO_2 uses an average extraction temperature of about 20°C. The use of low temperatures makes it suitable for the extraction of volatile and thermally unstable compounds; (5) this method uses the least amount of organic solvents (organic modifiers) and is environmentally more efficient process than other methods; (6) it is also automatic and can be directly connected by chromatographic methods such as gas chromatography or chromatography with supercritical fluid and quantification of highly volatile extracted compounds.

The main drawback of the system is that optimizing conditions are not easy. There are several dependent conditions for optimization, including flow rate, pressure, temperature, extraction mode (static or dynamic or combined), modifier, extraction time, extract collection system. In addition, SC-CO_2 is a matrix-dependent technique, i.e., the optimization conditions are highly dependent on the type of sample network and it must be carefully optimized for each type of sample network separately. Another problem with SC-CO_2 is that it has a high affinity not only for essential oil compounds but also for many other categories of low-density compounds found in vegetables, such as skin vaccines, fatty acids, pigments and resins. In one extraction step, simultaneous extraction of dermal vaccines is unavoidable because they are soluble in SC-CO_2 and are well-extracted; however, it is possible to obtain pure essential oils by extracting SC-CO_2 by choosing a complex process design. Also, extraction with SC-CO_2 requires the plant to be dried and at an additional cost. Other disadvantages of using SC-CO_2 are the difficult operating conditions and the high economic cost of the process.

1.5.4.3 Research Conducted by SFE Method

Extraction of medicinal substances from plants and natural resources using supercritical fluid has been considered by researchers over the years and many medicinal compounds of plant origin have been obtained from this method. Comparison between extraction with supercritical CO_2 (SC-CO_2) and water vapor distillation and Soxhlet for different essential oils has been reported.

Sun and Temelli (2006) extracted carotenoids from carrots by SFE method with canola oil as a co-solvent [25]. The results showed that not only hydrocarbon compounds such as alpha and beta carotene were recovered with supercritical CO_2 but also oxygenated carotenoids such as lutein were extracted. They said that using canola oil as an auxiliary solvent would increase the efficiency of carotenoid extraction. Zaidul et al. (2007) extracted palm kernel oil using SC-CO_2 and observed that the solubility of palm kernel oil increased at pressures above 20 MPa and temperatures above 40°C [26]. They reported the highest yields at 48.3 MPa and 80°C. Also, optimization of mannitol extraction from olive leaves was performed such that the yield of mannitol extracted by this method was lower than the value obtained by Soxhlet method.

Mohammadi et al. (2011) compared the compositions of coriander essential oil obtained from supercritical extraction with steam distillation [27]. The results showed that the extracted essential oil with supercritical CO_2 contained 29 compounds while in steam distillation extraction, 35 compounds were identified. This may be due to hydrolysis or oxidation of some compounds during steam distillation. In addition, the essential oil extracted with supercritical CO_2 was richer in the amount of linalool and limonene (66% and 3%) than the values obtained by distillation (57% and 0.02%).

1.5.5 Subcritical Water Extraction (SWE)

SWE is an extraction with subcritical water as a solvent with unique properties due to its strong hydrogen bonds in its structure. High boiling point, high dielectric constant and high polarity are other properties of water, which can only be used in the extraction of polar materials. Studies by various researchers have shown that by increasing the temperature from 25–250°C and applying a certain pressure, similar changes in water polarity, surface tension and viscosity occur close to a group of common mixtures of methanol or acetonitrile with water. Therefore, it can solve a wide range of medium- and low-polarity analytics. An important advantage of SWE is the reduction in the consumption of organic solvents. In addition, water is readily available and non-toxic and can be recycled or disposed of with minimal environmental problems. Hence, SWE has consistently become an efficient and low-cost method for extracting polar organic compounds from environmental soils, sediments and plant materials [28–31].

1.5.5.1 Subcritical Water Extraction Equipment

Figure 1.11 shows the laboratory equipment used in the preparation of natural essential oils by extracting subcritical water [32].

First, nitrogen gas is passed through distilled water for a period of time to remove oxygen from the solution. The water is then pumped at a constant flow rate by means of a precision pump through a needle valve to a temperature-balanced coil with a small diameter and a long length and then extracted into the cell. The cell is placed vertically inside the oven and water is pressurized from above, so that all the extracted material is immediately removed from the cell. The outlet of the extraction tube is connected by a tube to the needle valve which is located outside the wall of the oven chamber. This valve can be used as a flow control valve during tuber filling operations. A system is also required to generate and maintain pressure to keep water liquid at all temperatures used. Pressure-generating pumps or any other suitable system can be used for this purpose. The exit tube of the extraction cell leads to a sample collection vessel. Because hot water can cause some volatiles to be lost, a cooling coil is usually used after the needle outlet valve.

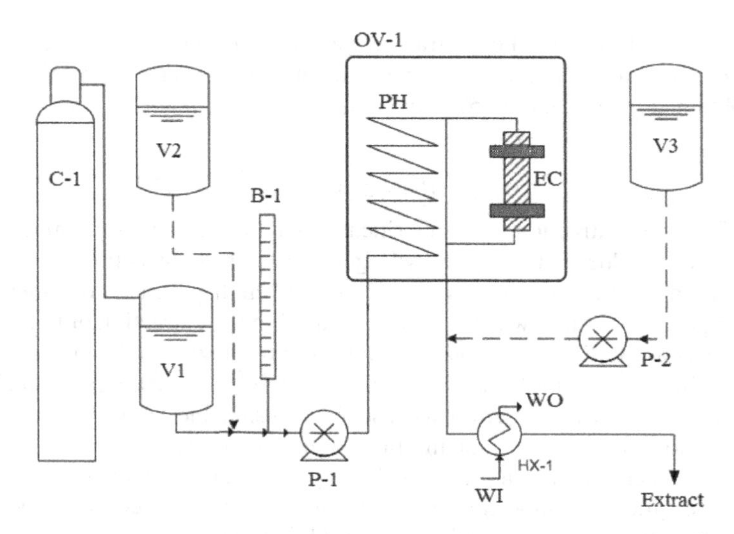

Figure 1.11 Extraction device with subcritical water in the preparation of natural essential oils V1, V2, V3 water tank; organic solvent and detergent; B1 burette; P1, P2 pump; OV1 oven; pH preheating; EC extraction tube; HX1 heat exchanger.

(Source: Khajenoori, M.; Haghighi Asl, A. and Eikani, M.H. (2014). *Ph.D. Thesis*).

1.5.5.2 Advantages and Disadvantages of SWE

Two inevitable consequences of using SWE to separate essential oils from plants is the reactive nature of water in these conditions, which is detrimental to the extracted analyte but can be obtained by using purer water and separating gases from water to prevent the extraction from this loss, and another is the relatively high temperature required, which necessitates early studies on the thermal stability of the extracted compounds. However, despite these disadvantages, this method has advantages such as (1) SWE method is fast and requires short extraction time; (2) the total amount of oxygenated compounds that can be extracted in this way is much higher than water distillation, solvent extraction and SFE, because water is the subcritical temperature of polar solvents; (3) SWE is an effective way to obtain aromatic compounds from plants while heavy monoterpenes and hydrocarbons and lipids remain; (4) the nature of the compounds extracted from SWE can only be controlled by increasing the temperature under pressure high enough to maintain the liquid state, allowing for high selectivity and the extraction of a wide range of species; (5) water is cheap and available; (6) does not require the drying stage of the plant; (7) perhaps the most important advantage of using subcritical temperature water is the reduction in the use of organic solvents.

1.5.5.3 Research Done by SWE Method

The first reports on the use of subcritical water for the extraction of active ingredients of plants date back to 1998, when the results of rosemary extraction were published. Extraction of essential oil from rosemary in subcritical water at 150°C showed that subcritical water tended to extract oxygenated substances and had a higher efficiency than steam distillation. Eikani et al. (2007) in their studies on the extraction of coriander seed essential oil with subcritical water found that temperature of 125°C, particle size of 0.5 mm, flow rate of 0.2 mL/min, pressure 20 bar and the duration of 120 min has led to an increase in the efficiency of aromatic oxygenated compounds, especially linalool, and a decrease in the efficiency of hydrocarbons compared to Soxhlet extraction by hexane and also extraction by water [33]. However, in this study, the highest overall efficiency was obtained for the water extraction method.

Extraction of lavender plant with subcritical water and in optimal conditions at 150°C, 20 bar and 3 mL/min, more oxygenated compounds were obtained by the extraction methods by water and Soxhlet by hexane [34]. Oxygen-free hydrocarbon components were not extracted with subcritical water and as a result, the obtained essential oil was better in terms of quality than the other two methods. The overall yield of essential oil was almost the same for all three methods, but the extraction time with subcritical water was much less reported than the other two methods. In 2009, Khajenooi et al. tested the SWE process with two common methods of water distillation and Soxhlet for *Zataria moltiflura Boiss* [28]. Based on the results of SWE, the highest essential oil yield was obtained at 150°C, flow rate of 2 mL/min, particle size of 0.5 mm and pressure of 20 bar for 150 min. A comparison was made between the three methods used in terms of speed, extraction efficiency and greenness of the method and the probability of combining the percentage of product. The results showed that the SWE is faster than the other two conventional methods. Extraction in this method lasted 150 min, while the distillation and Soxhlet methods lasted 3 h and 3.5 h, respectively. Extraction efficiency in subcritical water method was slightly higher than the other two methods. Comparison of different extraction methods with SWE is summarized in the Table 1.1 shown.

TABLE 1.1
Comparison of Extraction with Subcritical Water and Other Extraction Methods

Method and Time	Solvent	Solvent Amount	Advantages	Disadvantages
SWE 5–30 Min	Water	Very small volume of solvent	• Short extraction time • cheaper production cost • High efficiency • Contains less terpene Compounds and more oxygenated compounds • Better quality of essential oils • No simultaneous extraction of skin vaccines • No need to dry the plant • Possibility of selective extraction of compounds by temperature regulation	• Unavailability of commercial equipment • Degradation of active chemical compounds at high temperatures • Extremely high selectivity is useful for extracting oxygen-containing terpenes while oxygen-free terpenes are difficult to detect.
Soxhlet 4–48 h	Hexane, acetone dichloromethane, toluene, methanol	200–600 mL	• Ability to extract desirable volatiles (lipids have limited solubility in the solvent). • Solvent recovery • Shifting the equilibrium state (by repeatedly adding fresh solvent)	• Long extraction time • Possibility of thermal decomposition of stored compounds (extraction at the boiling point of solvent for a long time) • Simultaneous extraction of non-volatile compounds

(Continued)

TABLE 1.1 (*Continued*)
Comparison of Extraction with Subcritical Water and Other Extraction Methods

Method and Time	Solvent	Solvent Amount	Advantages	Disadvantages
			• Establish a relatively high extraction temperature by heating the distillation balloon • No need for filtration after leaching • Very simple and cheap	• Loss of volatile compounds in concentrating (using large volumes of organic solvents) • Lack of disturbance (stirring) in the machine to speed up the process
SFE 30 min–2 h	Carbon dioxide (sometimes with another solvent such as ethanol, methanol)	8–50 mL	• Ability to do it automatically • Low temperature • No solvent residue • High-efficiency CO_2 is cheaper and more abundant than organic solvents • Possibility of selectivity by changing the solvent density • It dissolves solids like a liquid solvent, but has a gas-like diffusion power and penetrates easily through solids.	• Use of organic solvents (modifiers) along with CO_2 to reduce polar constraints • Simultaneous extraction of waxes • High affinity for CO_2 with essential oil compounds • Need to dry the plant and spend extra money • High cost of purchase, repair and maintenance of devices

(*Continued*)

TABLE 1.1 (*Continued*)
Comparison of Extraction with Subcritical Water and Other Extraction Methods

Method and Time	Solvent	Solvent Amount	Advantages	Disadvantages
PLE 2–120 min	Hexane, acetone, dichloromethane, methanol	15–40 mL	• Perform the process completely automatically • Faster extraction time and lower solvent consumption than Soxhlet and ultrasound methods • Final extract is clean enough for direct analysis by GC/MS	• High price of equipment • Need for additional purification stage • Very high temperature and consequently very low monoterpene efficiency 5 times less recycled (limonene) than UAE and Soxhlet extraction • Simultaneous extraction of non-volatile species
MAE 30–60 min	Hexane, acetone	25–50 mL	• Extraction of 14 tanks simultaneously • Automate the process • Short extraction time • Suitable for thermally unstable species due to the use of low temperatures • High extraction efficiency. Considering the practical and economic aspects • Simplicity and low process cost	• The use of high microwave energies can lead to isomerization or degradation of the compound. • Need a filter or centrifuge to remove solid residue during the process • Low efficiency of non-polar and volatile compounds

(*Continued*)

TABLE 1.1 (*Continued*)
Comparison of Extraction with Subcritical Water and Other Extraction Methods

Method and Time	Solvent	Solvent Amount	Advantages	Disadvantages
Hydro-distillation 2–3 h	Water	1000–2000 mL	• No solvent residue • No need to smooth	• Potential loss of more polar terpenes and more active chemical compounds • Loss of volatile compounds • Low efficiency • Long extraction time
UAE 10–30 min	Hexane-acetone, acetone dichloromethane, toluene, methanol	< 50 mL	• High performance • Compared to conventional methods, cheap, simple and efficient • Increase extraction efficiency • Synthetics are faster • Possibility of testing at low temperature and possibility of extracting thermally unstable compounds	• Formation of free radicals and as a result, the potential for change in the constituent molecules

1.6 CONCLUSION

By observing the continuous increase in world demand for improving the technical steps of obtaining and analyzing materials using sustainable methods, it is possible to make predictions about the needs of laboratories in the future. Depending on the solvent used, consumption reduction and elimination should also be considered, as well as the possibility of replacing organic or toxic solvents. For each extraction method, it is necessary to produce valid data fast, with minimal operator intervention and cost-effectiveness. Also provide safety considerations for the operator, other employees in the operating environment. Innovative and environmentally friendly techniques in the extraction of various materials, which usually use less solvent and energy, such as the UAE, SFE, MAE, DIC and SWE are currently a dynamically developing area in research and applied industries. However, so far, there are only a few reports that point to the acceleration of the extraction process by combining these techniques. The main advantage of using hybrid techniques for extraction is to increase production efficiency and help protect the environment by reducing the use of solvents, fossil fuels and the production of hazardous materials. Currently, among the new methods, SWE and SFE are better processes. However, there is much research to improve the understanding of the extraction mechanism, remove technical barriers, to improve the design and scaling of extraction systems with these methods. The use of SWE as an alternative to organic solvents in various processes is still in its infancy and research.

REFERENCES

1. Pingret, D., Fabiano-Tixier, A.S. and Chemat, F. 2012. Accelerated methods for sample preparation in food. *Comprehensive Sampling and Sample Preparation*, 4: 446, 458.
2. Handa, S.S. 2008. Extraction technologies for medicinal and aromatic plants, Chapter 1, *An overview of extraction techniques for medicinal and aromatic plants*, Earth, Environmental and Marine Sciences and Technologies, Italy, 36–39.
3. Chemat, F., Abert-Vian, M., Fabiano-Tixier et al. 2019. Green extraction of natural products: Origins, current status, and future challenges. *Trends in Analytical Chemistry*, 118: 248–263.
4. Gedye, R., Smith, F., Westaway, K. et al. 1986. The use of microwave ovens for rapid organic synthesis. *Tetrahedron Letters*, 27 (3): 279–282.

5. Ganzler, K., Salgo, A. and Volko, K. 1986. Determination of lead contented on particulate matter filters by microwave extraction and analysis by atomic absorption spectrometry. *Journal of Chromatography*, 371: 299–306.
6. Lucchesi, M.E., Chemat, F. and Smadja, J. 2004. Solvent-free microwave extraction of essential oil from aromatic herbs: comparison with conventional hydro-distillation. *Journal of Chromatography A*, 1043: 323–327.
7. Hemwimon, S., Pavasant, P. and Shotipruk, A. 2007. Microwave-assisted extraction of antioxidative anthraquinones from roots of *Morinda citrifolia. Separation and Purification Technology*, 54(1): 44–50.
8. Pérez-Serradilla, J.A., Ortiz, M.C., Sarabia, L.A. et al. 2007. Focused microwave-assisted Soxhlet extraction of acorn oil for determination of the fatty acid profile by GC-MS comparison with conventional and standard methods. *Analytical and Bioanalytical Chemistry,* 388(2): 451–462.
9. Lianfu, Z. and Liu, Z. 2008. Optimization and comparison of ultrasound/microwave assisted extraction (UMAE) and ultrasonic assisted extraction (UAE) of lycopene from tomatoes. *Ultrasonics Sonochemistry*, 15(5): 731–737.
10. Bousbia N, Vian MA, Ferhat MA. et al. 2009. Comparison of two isolation methods for essential oil from rosemary leaves: Hydrodistillation and microwave hydrodiffusion and gravity. *Food Chemistry*, 114(1): 355–362.
11. Gujara, J.G., Wagha, S.J. and Gaikarb, V.G. 2010. Experimental and modeling studies on microwave-assisted extraction of thymol from seeds of *Trachyspermum ammi* (TA). *Separation and Purification Technology*, 70: 257–264.
12. Pingret, D., Fabiano-Tixier, A.S. and Chemat, F. 2012. *Accelerated methods for sample preparation in food: Comprehensive sampling and sample preparation.* Academic Press, Oxford, pp. 441–455.
13. Alirezapour, N., Haghighi Asl, A. and Khajenoori, M. 2020. Ultrasound-assisted extraction of Thymol from *Zataria multiflora Boiss.*: Optimization by response surface methodology and comparison with conventional Soxhlet extraction. *Bulgarian Chemical Communications*, 52(4): 419–427.
14. Hemwimol, S., Pavasant, P. and Shotipruk, A. 2006. Investigated the use of ultrasonic extraction to improve the extraction of anthraquinones from *Morinda citrifolia* root with solvent. *Ultrasonics sonochemistry*, 13(6): 543–548.
15. Martino, E., Ramaiola, I., Urbano, M. et al. 2006. Microwave-assisted extraction of coumarin and related compounds from *Melilotus officinalis* (L.) Pallas as an alternative to Soxhlet and

ultrasound-assisted extraction. *Journal of Chromatography A*, 1125: 147–151.
16. Jacques, R.A., Freitas L.S. Pérez, V.F. et al. 2007. The use of ultrasound in the extraction of *Ilex paraguariensis* leaves: A comparison with maceration. *Ultrasonics Sonochemistry*, 14: 6–12.
17. Wang, J., Suna, B., Cao, Y. et al. 2008. Optimisation of ultrasound-assisted extraction of phenolic compounds from wheat bran, *Food Chemistry*, 106: 804–810.
18. Kongkiatpaiboon, S. and Gritsanapan, W. 2012. HPLC quantitative analysis of insecticidal didehydrostemofoline and stemofoline in *Stemona collinsiae* root extracts, *Phytochemical Analysis*, 23(5), 554–558.
19. Allaf, K., and Vidal, P. 1989. Feasibility Study of a new process of drying/swelling by instantaneous decompression toward vacuum of in pieces vegetables in view of a rapid re-hydration. Gradient Activity Plotting University of Technology of Compiegne UTC N° CR/89/103, Industrial SILVA-LAON Partner.
20. Kristiawan, M., Sobolik, V. and Allaf, K. 2008. Isolation of *Indonesian Cananga* oil using multi-cycle pressure drop process. *Journal of Chromatography A*, 1192(2): 306–18.
21. Ben Amor, B., and Allaf, K. 2009. Impact of texturing using instant pressure drop treatment prior to solvent extraction of anthocyanins from Malaysian Roselle (*Hibiscus sabdariffa*). *Food Chemistry*, 115(3): 820–825.
22. Besombes, C., Berka-Zougali, B. and Allaf, K. 2010. Instant controlled pressure drop extraction of lavandin essential oils: Fundamentals and experimental studies. *Journal of Chromatography A*, 1217(44): 6807–6815.
23. Pedrosa, M.M., Cuadrado, C., Burbano, C. et al. 2012. Effect of instant controlled pressure drop on the oligosaccharides, inositol phosphates, trypsin inhibitors and lectins contents of different legumes. *Food Chemistry*, 131(3): 862–868.
24. Khaw, K.Y., Parat, M.O., Shaw, P.N. et al. 2017. Solvent supercritical fluid technologies to extract bioactive compounds from natural sources: A review. *Molecules*, 22: 1186–1208.
25. Sun, M. and Temelli, F. 2006. Supercritical carbon dioxide of carotenoids from carrot using canola oil as a continuous co-solvent. *Journal of Supercritical Fluids*, 37(3): 397–408.
26. Zaidul, I.S.M., Norulaini, N.A.N., Omar, A.K.M. et al. 2007. Separation of palm kernel oil from palm kernel with supercritical carbon dioxide using pressure swing technique. *Journal of Food Engineering*, 81(2): 419–428.
27. Mohammadi, S. and Saharkhiz, M.J. 2011. Changes in essential oil content and composition of catnip (*Nepeta cataria L.*) during

different developmental stages. *Journal of Essential Oil Bearing Plants*, 14 (4): 396–400.

28. Khajenoori, M., Haghighi Asl, A., Hormozi, F. et al. 2009. Subcritical water extraction of essential oils from *Zataria multiflora Boiss. Journal of Food Process Engineering*, 32(6): 804–816.
29. Khajenoori, M., Haghighi Asl, A. and Eikani, M.H. 2015. Subcritical water extraction of essential oils from *Trachyspermum ammi* seeds, *Journal of Essential Oil-Bearing Plants*, 18(5): 1165–1173.
30. Khajenoori, M., Haghighi Asl, A. and Eikani, M.H. 2015. Optimization of subcritical water extraction of *Pimpinella anisum* seeds. *Journal of Essential Oil-Bearing Plants*, 18(6): 1310–1320.
31. Haghighi Asl, A. and Khajenoori, M. 2013. Subcritical water extraction, *Mass transfer-advances in sustainable energy and environment oriented numerical modeling*, Intech, Rijeka.
32. Khajenoori, M., Haghighi Asl, A. and Eikani, M.H. 2014. *Investigation of thermosynthetic parameters of plant extraction with subcritical water.* Doctoral thesis, Semnan University, Iran.
33. Eikani, M.H., Golmohammad, F. and Roshanzamir, S. 2007. Subcritical water extraction of essential oils from coriander seeds (*Coriandrum sativumMill.*). *J. Food Eng*, 80(2): 735–740.
34. Eikani, M.H., Golmohammad, F., Shokrollahzadev, S. et al. 2006. Superheated water extraction of *Lavandula latifolia Medik* volatiles: comparison with conventional techniques. *Journal of Essential Oil-Bearing Plants*, 20: 482–487.

Chapter 2

REVIEW OF SUBCRITICAL WATER EXTRACTION (SWE)

2.1 INTRODUCTION

For centuries, humans have tried to separate, concentrate and purify aromatic substances in nature. Common methods of extracting essential oils include distillation with water vapor and extraction by water and organic solvent. Loss of some volatile materials, low extraction efficiency, decomposition of unsaturated compounds due to thermal effects and residual toxicity of the solvent in the extraction phase are some of the shortcomings of these methods. Another important issue in the essential oil industry is achieving the right quality. All of this has led to the use of new extraction methods, including the use of supercritical solvents in the extraction process, with most of the extraction focused on the use of carbon dioxide. This method also has limitations due to its high cost and the need for side operations such as drying raw materials.

A number of researchers in study of new and green solvents have used pressurized water at temperatures above 100°C and below critical temperatures (374°C), in which case water is referred to as "subcritical water". Due to the similarity of the name with the supercritical process, it is preferred in scientific sources to use the name of superheated water.

2.2 SUBCRITICAL WATER EXTRACTION (SWE)

In order to reduce the use of organic solvents, SWE is an optimal green solvent extraction method that leads to the use of pressurized water at high temperatures and controlled pressure conditions. Comparison of the major physical properties of water at different temperatures is shown in Table 2.1 [1–5]. Among the solvents used

TABLE 2.1
Physical and Chemical Properties of Water at Different Temperatures

	Ambient Temperature	Subcritical Water	Supercritical Water
Temperature [°C]	0–100	100–374	>374
Vapor pressure [MPa]	0.003 (24°C)	0.1 (100°C)–22.1 (374°C)	>22.1
Aggregate state	Liquid	Liquid	No phase separation
Density [gcm^{-3}]	0.997 (25°C)	0.958 (101°C, 0.11 MPa) 0.692 (330°C, 30 MPa)	Between gas-like and liquid-like densities, for example 0.252 (410°C, 30 MPa)
Viscosity [μPas]	L: 884 G: 9.9 (25°C)	L: 277 G: 12.3 (101°C)	Low
Heat capacity CP [$Jg^{-1}K^{-1}$]	L: 4.2 G: 2.0 (25°C)	L: 4.2 G: 2.1 (101°C) L: 69 G: 145 (371°C)	1300 (400°C, 25 MPa)
Compressibility	No	Slightly increased, but still a liquid (at 370°C)	Yes
Ion product K_w [mol^2L^{-2}]	10^{-14} (increasing to 10^{-12} at 100°C)	Increases from 10^{-12} (100°C) to 10^{-11} (300°C)	Strongly decreasing to below 10^{-20} (400°C) and below 10^{-23} (550°C); increases slightly with P

L = liquid phase, G = gas phase.
Source: Moller, M.; Nilges, P.; Harnisch, F. et al. *ChemSusChem* (2011), 4(5): 566–579.

in material extraction, water is a solvent with unique properties due to its strong hydrogen bonds in its structure. The presence of these bonds has turned water into a polar solvent with a high permeability coefficient at normal temperatures that can only be used to extract polar materials.

Studies by various researchers have shown that by increasing the temperature and applying a certain pressure, the polarity of water can change close to a group of alcohols. For example, at 220°C, the electrical conductivity of water is about 30, which is similar to the value of the electrical conductivity of methanol at room temperature (32) [6]. Therefore, it can solve a wide range of medium- and low-polarity analyzes. The dielectric constant of water at various temperatures is shown in Table 2.2. An important advantage of SWE is the reduction in the consumption of organic solvents. In addition, water is readily available and non-toxic, which can be recycled or disposed of with minimal environmental problems. Thus, SWE has consistently become an efficient and low-cost way to extract low-polar organic compounds from environmental soils, sediments and plant materials [7, 8].

The term "subcritical water" is used to mean the phase of water condensation phase between the temperature ranges of 100°C (water boiling point) to 374°C (water critical point). In other words, at any temperature, the system pressure is adjusted so that the water stays in

TABLE 2.2
Dielectric Constant of Water at Various Temperatures

Temperature (°C)	Pressure (bar)	Dielectric Constant
100	1.1	55.39
110	1.4	52.89
120	2.0	50.48
130	2.7	48.19
140	3.6	46.00
150	4.8	43.89
160	6.2	41.87
170	7.9	39.96
180	10.0	38.10
190	12.6	36.32
200	15.6	34.59
210	19.1	32.93
220	23.2	31.32
230	28.0	29.75
240	33.5	28.24

Source: Khajenoori, M., Haghighi Asl, A. and Eikani, M.H. (2014). *Ph.D. Thesis*, Semnan University, Iran.

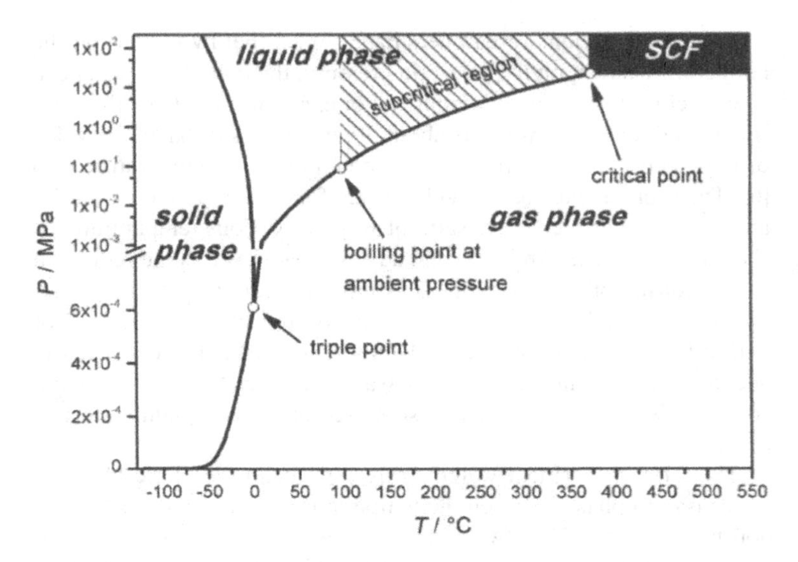

Figure 2.1 Phase diagram of water showing its aggregate states. Characteristic points as well as the subcritical region are indicated. SCF = supercritical fluid.

(Source: Moller, M.; Nilges, P.; Harnisch, F. et al. *ChemSusChem* 2011, 4(5): 566–579).

the liquid phase. This range can be clearly seen in the water temperature-pressure diagram (Figure 2.1) [5]. At these temperatures, water effectively dissolves low-polarity compounds, such as polycyclic aromatic hydrocarbons (PAHs) and polychlorinated biphenyl (PCBs) [9].

Other common terms have been used, such as "pressurized hot water", "almost critical water", "high temperature water" and "compressed hot water". The use of subcritical water as a liquid at high temperatures has been reported for the first time in the experiments of Howthorn et al. to extract some polar and non-polar analyzes of soil samples in 1994 [10].

It is obvious that if the pressure at any temperature is lower than the boiling point pressure at that temperature, superheated steam will be produced which will inevitably have a lower electrical conductivity than the liquid phase and the penetration rate will be the same as the penetration coefficient of the gas phase. As a result, a solvent with different properties will be produced with the SW considered in this method. It should also be noted that due to the increase in solubility

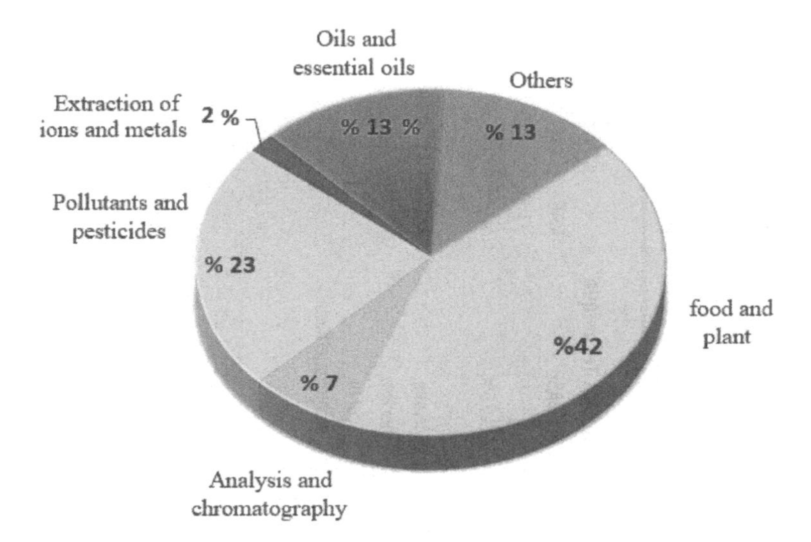

Figure 2.2 Graphic display of the results of the search of articles on the Scoops site, using the keywords "subcritical water extraction" and "Superheated water extraction" in 2000–2011.

of materials with temperature, with increasing temperature there is a possibility of side products in the products of extraction. Also, due to the high temperature, it is possible to destroy the structure of some materials, especially the materials available in plant and food matrix [11, 12].

Figure 2.2 shows the overall distribution of the progress made in the use of SW in various fields from 2000 to 2011, according to the Scoops website. The main use of SW: Foods and plants (42%), followed by pollutants and pesticides (23%).

The first reports of the use of SW for the extraction of plant active substances date back to 1997–98, when the results of the rosemary extraction were published [13]. Since then, several scientific studies on the use of this method in extracting essential oils have been published each year.

2.3 COMPARISON OF SWE AND OTHER METHODS

Comparison of SWE and other methods is shown in Table 2.3. To compare different extraction methods, different factors such as yield, extraction time, method selectivity, solvent type, amount of solvent

TABLE 2.3
Comparison of Soxhlet Extraction, SWE, Supercritical Fluid Extraction (SFE), Pressurized Liquid Extraction (PLE) and Microwave Assisted Extraction (MAE)

	Traditional Technique	**Instrumental Extraction Technique**			
Extraction technique	Soxhlet	SWE	SFE	PLE=ASE	MAE
Typical extraction time	4–48 h	5–30 min	30–90 min	12–20 min	30–60 min
Typical solvent	Acetone-hexane, acetone-dichloromethane, dichloromethane, toluene, methanol	Water	CO_2/CO_2+modifier	Acetone hexane, acetone-dichloromethane	Acetone hexane
Typical solvent consumption (ml)	300	- /a few milliliters for elution of analytes	8–50/no solvent needed in on-line SFE-GC	15–40	25–50
Selectivity for compound classes	Non-selective	Selective	Lightly selective	Non-selective	Non-selective

(*Continued*)

TABLE 2.3 (*Continued*)
Comparison of Soxhlet Extraction, SWE, Supercritical Fluid Extraction (SFE), Pressurized Liquid Extraction (PLE) and Microwave Assisted Extraction (MAE)

	Traditional Technique		**Instrumental Extraction Technique**		
Selectivity for sample matrix	Some selectivity	Selective	Selective	Non-selective	Non-selective
Benefits	Simple well-known procedure, easy to carry out, cheaper production cost, also automated	No organic solvent needed; wet samples can be extracted without drying	No or little organic solvent needed, also automated	Fully automated	Generally, 14 vessels extracted simultaneously, also automated
Disadvantages	Time consuming, a lot of manual work, large consumption of organic solvent	No commercial equipment available (ASE, SFE and MAE equipment applicable in some cases), high temperature	Expensive equipment, need for modifiers when CO_2 the extractant	Expensive equipment, blockages, frequent need for additional clean-up	Need for additional clean-up to remove the matrix

Source: Andersson, T.; *Doctoral thesis* (2007), University of Helsinki, Helsinki.

required, energy required for extraction, economic aspects as well as environmental aspects can be considered [14–16].

The two inevitable consequences of using SW to separate essential oils from plants are the reactive nature of water in these conditions, which is harmful to the extracted analyte, but can be achieved by using purer water and separating gases from water for execution prevented this loss, and another requires a relatively high temperature, which requires preliminary studies of the thermal stability of the extracted compounds. However, this method has the disadvantages but it has the benefits in return with carbon dioxide extraction. In this method, the extraction of skin vaccines at the same time is not available in a single extraction stage. Therefore, it is not necessary to use a complex system that is required in the supercritical carbon dioxide (SC-CO_2) extraction method to obtain purer essential oils.

In extraction with SC-CO_2, the plant often needs to be dried because the water is not very soluble in it and tends to form undesirable compounds. Therefore, in the SWE method, the drying stage is eliminated, which causes additional costs and the risk of wasting fragrant volatile compounds in the extraction. Due to the much lower pressure required, SWE equipment is much cheaper than SC-CO_2 equipment. The use of water provides significant savings in maintenance costs, as the cost of the SC-CO_2 is extremely high. The final yield of three major compounds of clove essential oil in extracting Soxhlet for 24 hours with dichloromethane is slightly less than SWE at 20 bar and 150°C and 2 mL/min and only 100 min [17]. Therefore, SWE is the fastest method and prevents the presence of toxic organic residues in extracted materials. In addition, the nature of the compounds extracted by SWE can only be controlled by increasing the temperature at a sufficiently high pressure to maintain the liquid state, allowing high selectivity and a wide range of species to be extracted. Therefore, SWE has been proposed as a strong and bright future against both conventional methods (water vapor distillation, solvent extraction) and new methods (SC-CO_2 extraction) and has significant definite advantages over them.

2.4 APPLICATIONS OF SWE

SWE is mainly used for solid and powder samples and, as these matrices are more compatible with the extraction system process. SWE method has been used successfully on foods and plants to extract flavors and aromas, as well as their biologically active compounds [18]. In addition, SWE method is also used in the extraction of organic contaminants from food for food safety analysis [19]. Changes in the physical and chemical properties of water provides the extraction of non-polar organics such as polycyclic aromatic hydrocarbons (PAHs) and polychlorinated biphenyls (PCBs) from soil and sedimentary environmental samples [20]. SWE, due to the green nature and its feasibility to extract a wide range of compounds under SW conditions, is considered in biodegradable processes with recovery of pesticides and herbicides from soil/sediments [21].

2.4.1 Extraction of Biologically Active and Nutritional Compounds from Plant and Food Materials

In recent years, SWE has gradually become a useful option for isolating biologically active and nutritional compounds from plants. SWE is a direct, non-refining method for recovering analyzes. This reduces costs, so that the extraction analyzes are safe for further testing, processing and human consumption. The purpose of extracting the target compounds from *Gastrodia elata* and *Stevia rebaudiana* using SWE is comparable or higher than reflux under water heating [22]. Studies have been conducted on the concentration of plant substances for the extraction of biologically active substances as well as the main volatile essential oils in optimal extraction conditions. In this section, the extracted materials are divided into three groups of antioxidants, phenolic compounds and essential oils and are presented in three separate sections.

2.4.1.1 Antioxidants

Antioxidants have many uses due to their beneficial properties for human health and preventing the oxidation of fats. Extraction of plant materials by traditional methods such as distillation or solid-liquid extraction requires the use of organic solvents. It is possible that these solvents will remain in the final product. Due to the need of food industries for antioxidants produced by safe and environmentally friendly methods, researchers have turned their attention to these efficient methods such as extraction with SW and supercritical fluids. Many cases of antioxidant extraction of plants using SW have been investigated.

2.4.1.2 Phenolic Compounds

Phenolic compounds are powerful antioxidants and free radical scavengers that prevent the oxidation of fats and low-density lipoproteins, as well as reduce the risk of heart disease. In addition, they have antibacterial and antifungal properties. Therefore, extracting them from solid foods in a suitable way is the first step in analyzing these compounds. Anthocyanins are natural, non-toxic, water-soluble pigments that cause red, purple, blue and orange colors of fruits, plants and flowers. Extraction of catechins and proanthocyanidins from dry grape seeds is possible compared to normal extraction with 75% methanol [23].

2.4.1.3 Essential Oils

Essential oils are volatile, reflective liquids and aromatic and colorless compounds of terpene and alcoholic origin. Essential oils are a mixture of different substances with very different chemical compounds that cause a pleasant smell or taste in the plant. Essential oils are found in many plants, the most important of which contain essential oils, such as mint, umbrella, pine, bay leaf, and some plants of the rosacea and chicory family.

Essential oils accumulate in some plant tissues, such as the center of the cell, or in the storage of essential oils under the hair follicles, small glands or in the intercellular space. These materials oxidize and polymerize rapidly to resin at temperatures above 50 to 60°C and in the presence of air (exposed to light). Therefore, they become known as volatile oils, aromatic oils and oils. They are called ester oils and so on.

Essential oils are now extracted from natural products using various solvents or distilled water. Loss of some of the volatile compounds, low efficiency of the extraction process and decomposition of unsaturated compounds due to heat or hydrolysis and residual presence of toxic solvents in the extraction product are some of the potential problems of these extraction methods. One of the methods that has recently been studied for extracting essential oils from different plants is the method of SWE [24].

Andres Aman et al. have extracted three combinations of desert mint by extraction with SW, steam distillation and extraction with SC-CO_2. According to them, extraction with SW at 150°C and pressure 125 MPa is more efficient than extraction with supercritical fluid. Comparing the three methods under the same conditions also shows that the efficiency of the distillation process is higher than the other two processes [25]. Many cases of essential oil extraction using SW have been studied [26].

2.4.1.4 Other Plant and Food Ingredients

SWE is also a common method for extracting compounds from other plant materials. The stability of these compounds at high temperatures and their extraction efficiency has been studied in comparison with other extraction methods. Using SWE, five different capsaicinoids (hydrocapsaicin, capsaicin, dihydrocapsaicin, isomer dihydrocapsaicin and hemodia hydroxypsycin) were successfully detected at 200°C, and were measured by HPLC instrument before reducing their efficiency at higher temperatures [27].

2.4.2 Eliminate Organic Contaminants from Food

In recent years, the chemical contamination of food contaminants has grown significantly. These chemical contaminants can be classified into four main categories: (1) Pesticides, (2) veterinary drugs, (3) chronic environmental chemicals and (4) naturally occurring toxicants [28]. Based on the use of the Soxhlet method, the extraction of contaminants from food is usually accompanied by long-term extraction and treatment methods. These methods are laborious and time consuming and usually consume a large amount of toxic organic solvents.

2.5 ECONOMIC STUDY OF SUBCRITICAL WATER EXTRACTION

SWE requires 305 kJ/kg energy to heat water from 30°C to 150°C at a constant pressure of 15 bar. In contrast, steam distillation, which is one of the most common methods of extracting essential oils, requires 2550 kJ/kg energy to convert water at 30°C to steam at 100°C. In addition, the amount of heat in the SWE can be returned, but in steam distillation, a small amount of it is recovered. The amount that can be recycled depends on the size of the heat exchangers. For example, if a difference of 30°C in the temperature of the currents in the heat exchanger is designed, 75% of the heat can be recovered. In steam distillation, condenser cooling water can be used in the boiler, but the maximum temperature can probably be 70°C. This means that only 6.5% of the heat can be recovered. The thermal efficiency of SWE is about 20 times per kilogram of SW compared to steam distillation. SWE of 1 kg rosemary with 10 kg of water, by steam distillation with a weight ratio of 1:1 have the same efficiency. Although in the SWE, the required amount of water is 10 times more, but the thermal superiority of this method is preferred. In almost all reports, extraction speed with SW is reported to be faster than conventional methods, and this method requires less time and less manpower.

REFERENCES

1. NIST. 2008. Thermophysical Properties of Fluid Systems, U.S. Secretary of Commerce on behalf of the United States of America, http://webbook.nist.gov/chemistry/fluid.
2. Kruse, A. and Gawlik, A. 2003. Biomass conversion in water at 330–410°C and 30–50 MPa. identification of key compounds for indicating different chemical reaction pathways. *Industrial & Engineering Chemistry Research*, 42: 267–279.
3. Krammer, P. and Vogel, H., J. 2000. Hydrolysis of esters in subcritical and supercritical water. *Supercritical Fluids*, 16: 189–206.
4. Meyer, A., McClintock, R.B., Silvestri, G.J. et al. 1992. *Spencer steam tables: Thermodynamic and transport version 6*, ASME, New York.
5. Moller, M., Nilges, P., Harnisch, F. and Schroder, U. 2011. Subcritical water as reaction environment: fundamentals of hydrothermal biomass transformation. *ChemSusChem*, 4(5): 566–579.
6. Haghighi Asl, A. and Khajenoori, M. 2013. Subcritical water extraction, *Mass transfer-advances in sustainable energy and environment oriented numerical modeling*, Intech, Rijeka.
7. Lindholm-Lehto, P.C., Ahkola, H.S.J. and Knuutinen, J.S. 2017. Procedures of determining organic trace compounds in municipal sewage sludge-a review. *Environmental Science and Pollution Research*, 24: 4383–4412.
8. Khajenoori, M., Haghighi Asl, A. and Eikani, M.H. 2014. Investigation of thermosynthetic parameters of plant extraction with subcritical water. *Ph.D. Thesis*, Semnan University, Iran.
9. Yang, Y., Bowadt, S., Hawthorne, S.B. et al. 1995. Subcritical water extraction of polychlorinated biphenyls from soil and sediment. *Analytical Chemistry*, 67(24): 4571–4576.
10. Hawthorne, S. B., Yang, Y., and Miller, D. J. 1994. Extraction of organic pollutants form environmental solids with sub-and supercritical water, *Analytical Chemistry*, 66: 2912–2920.
11. Pineiro, Z., Palma, M. and Barroso, C.G. 2004. Determination of catechins by means of extraction with pressurized liquids. *Journal of Chromatography A*, 1026 (1–2): 19–23.

12. Khajenoori, M, Haghighi Asl, A. and Hormozi, F. et al. 2009. Subcritical water extraction of essential oils from *Zataria multiflora Boiss. Journal of Food Process Engineering*, 32(6): 804–816.
13. Basile, A., Jimenez-Carmona, M.M. and Clifford, A.A. 1998. Extraction of rosemary by superheated water. *Journal of Agricultural and Food Chemistry*, 46(12): 5205–5209.
14. Hyotylainen, T. 2009. Critical evaluation of sample pretreatment techniques. *Analytical and Bioanalytical Chemistry*, 394: 743–758.
15. Ong, E.S., Han Cheong, J.S. and Goh, D. 2006. Pressurized hot water extraction of bioactive or marker compounds in botanicals and medicinal plant materials. *Journal of Chromatography A*, 1112: 92–102.
16. Andersson, T. 2007. Parameters Affecting the Extraction of Polycyclic Aromatic Hydrocarbons with Pressurised Hot Water. Doctoral thesis, University of Helsinki, Helsinki.
17. Clifford, A.A., Basile, A. and Al-Saidi, S.H.R. 1999. A comparison of the extraction of clove buds with supercritical carbon dioxide and superheated water. *Fresenius Journal of Analytical Chemistry*, 364: 635–637.
18. Mukhopadhyay, M. and Panja, P. 2008. A novel process for extraction of natural sweetener from licorice (Glycyrrhiza glabra) roots. *Separation and Purification Technology*, 63: 539–545.
19. Marchese, S., Perret, D., Bafile, E. et al. 2009. Pressurized liquid extraction coupled with LC-ESI-MS-MS for the determination of herbicides chlormequat and mepiquat in flours. *Chromatographia*, 70: 761–767.
20. Itoh, N., Numata, M., Aoyagi, Y. et al. 2008. Comparison of low-level polycyclic aromatic hydrocarbons in sediment revealed by Soxhlet extraction, microwave-assisted extraction, and pressurized liquid extraction. *Analytica Chimica Acta*, 612: 44–52.
21. Kronholm, J., Kalpala, J., Hartonen, K. et al. 2002. Pressurized hot water extraction coupled with supercritical water oxidation in remediation of sand and soil containing PAHs. *The Journal of Supercritical Fluids*, 23: 123–134.
22. Smith, R.M. 2002. Extractions with superheated water. *Journal of Chromatography A*, 975: 31–46.
23. Marino, M.G., Gonzalo, J.C.R., Ibanez, E., Moreno, C.G. 2006. Recovery of catechins and proanthocyanidins from winery by-products using subcritical water extraction. *Analytica Chimica Acta*, 563: 44–50.
24. Kim, J.W., Nagaoka, T., Ishida, Y. et al. 2009. *Separation and Purification Technology*, 44: 2598–2608.
25. Jimenez-Carmona, M.M., Ubera, J.L. and Luque de Castro, M.D. 1999. Comparison of continuous subcritical water extraction

and hydrodistillation of marjoram essential oil. *Journal of Chromatography A*. 855: 625–632.
26. Khajenoori, M., Haghighi Asl, A. and Eikani, M.H. 2015. Optimization of subcritical water extraction of *Pimpinella anisum* seeds. *Journal of Essential Oil-Bearing Plants* 18(6): 1310–1320.
27. Teo, C.C., Tan, S.N., Yong, J.W.H. et al. 2009. Validation of green-solvent extraction combined with chromatographic chemical fingerprint to evaluate quality of *Stevia rebaudiana Bertoni*. *Journal of Separation Science*, 32: 613–622.
28. Bogialli, S. and Corcia, A. 2007. Matrix solid-phase dispersion as a valuable tool for extracting contaminants from foodstuffs. *Journal of Biochemical and Biophysical Methods*, 70: 163–179.

Chapter 3

SOLUBILITY OF SUBCRITICAL WATER

3.1 INTRODUCTION

The design and development of the extraction process with subcritical water (SW) depends on knowing the solubility of the SW. A good solubility accelerates the initial stages of extraction and somewhat reduces the time it takes to complete the process [1]. Therefore, the solubility rate is one of the important parameters for obtaining the appropriate extraction rate under optimal operating conditions [2–4]. Solubility data provides very important information about the efficiency and time of extraction and the optimal extraction temperature. Therefore, it is necessary to study the parameters that affect the solubility in SW. This data is also very important when scale is increased. In order to optimize the extraction process with SW, it is necessary to know the solubility of the compounds in SW. Construction of pilot and industrialization of the extraction process requires the solubility of the compounds. For this purpose, a better understanding of the thermodynamics of the solubility of compounds in SW is necessary, such that thermodynamic modeling and simulation of solubility for equipment design is inevitable.

Due to the limitations of solubility measurement experiments in SW, today the modeling of solubility and modification of existing models for obtaining more suitable models has been widely considered. These models are needed to predict the solubility of materials in SW before laboratory work is performed. Also, these solubility models are very important when the process becomes semi-industrial or industrial scale. Explaining the appropriate phase behavior of SW and solute is essential for selecting the optimal extraction temperature. Studies that focus on predictive methods of solubility of soluble matter in SW are very limited. Therefore, due to the lack of research resources, more detailed studies and researches are needed in this field.

The solubility of an organic compound in SW is often several degrees greater than its solubility in water at normal temperatures. There are two reasons for this. (1) In general, increasing entropy increases temperature solubility. (2) As the water temperature rises, its structure changes due to the failure of the hydrogen bond structure and its polarity decreases. The degree of liquid separability, meaning the electrical conductivity (or dielectric constant), is much lower for SW than water at normal temperature and can dissolve organic compounds, especially if they can be polarized or be a little polar.

Rising temperatures affect intermolecular forces. Intermolecular forces between solvent and soluble molecules in a layer around soluble molecules have a significant effect on solubility [5]. Since the solubility of the component plays a key role in the extraction processes, any factor that increases the solubility improves the efficiency of the extraction process. It should be noted that if the components are unstable at high temperatures, the polarity of the water can be reduced by adding another solvent such as ethanol, and as a result, the solubility action can be performed at lower temperatures [6]. Two major methods for measuring the solubility of compounds in water below critical temperatures have been reported in the papers. The first is the static method, in which the solubility of the compound is measured in a closed cell, and the second is the dynamic method, in which the solvent enters the system with a high-pressure pump, certainly with a sufficiently low intensity to maintain equilibrium.

3.2 EFFECTIVE PARAMETERS ON SOLUBILITY

3.2.1 Solvent Type

The thermodynamic properties of water (under environmental conditions and SW) are expressed in terms of hydrogen bonding strength and hydrogen bonding structure [6]. The change in hydrogen bond strength is reflected in the amount of dielectric constant and the amount of heat of evaporation. At lower temperatures, the hydrogen bond is stronger and the dielectric constant is higher [7]. As the water temperature rises, the hydrogen bonding strength decreases, leading to a decrease in the dielectric constant value [8]. Decreasing the hydrogen bonding strength and thus reducing the water polarity leads to an increase in the solubility of the hydrophobic organic compounds in it, and the water in these conditions acts like non-polar solvents.

3.2.2 Temperature

The solubility of hydrophobic organic compounds increases with increasing temperature [9]. In other words, it can be said that with increasing temperature, better mass transfer takes place. On the other hand, raising the temperature may cause problems such as corrosion, exacerbation of reactions such as hydrolysis and vaccination and destruction of the sample, especially in plant tissues and food.

3.2.3 Flow Rate

In the very low flow rate range (less than 1 mL/min), the higher the flow rate, the higher the apparent velocity and therefore the higher the Reynolds number. Considering the relationship between the Sherwood number and Reynolds number, the higher the Reynolds number, the higher the Sherwood number, and as a result, the mass transfer coefficient increases. Therefore, more mass transfer will increase the rate of dissolution in SW.

3.2.4 Pressure

Pressure has little effect on the solubility of hydrophobic organic compounds in SW. For example, when the pressure increases, the solubility of organic compounds decreases slightly [10]. For example, the solubility of carboxylic acids decreases in the pressure range above 350 bar [11], while the solubility of anthracene in water in the pressure range of 2800 bar decreases by one degree [3]. SWE and other SW processes are usually performed at pressures between 20 bar and 100 bar, which solubility changes can be ignored with pressure changes [12].

3.2.5 Dynamic or Static Mode

Determination of SW solubility can be done in both static and dynamic modes. Similar devices are used for both dynamic and static modes using SW as solvent. However, in each case, the efficiency depends on different process variables. In dynamic mode, the time to achieve equilibrium depends on water temperature, water flow rate and polarity of the analysis [13]. In determining the SW solubility, high-pressure water passes through a narrow chamber. However, prolonged heating may cause the compound to decompose, so optimizing the time to reach equilibrium is very important. While the dynamic method is usually faster than the static method due to the constant presence of fresh solvent, it is likely that more water will be used to collect the material and the collected product will be thinner. Flow rate of 1 or 1.5 mL/min is commonly used in the dynamic method. However, higher flow rates will usually improve the efficiency of very concentrated samples, as the total volume of water as well as the physical mass transfer of analyzes from the matrix will increase [14]. In the static method, the efficiency depends on the matrix equilibrium constant and the solubility of the compounds at high temperatures. Therefore, the low volume of water used may lead to incomplete results in highly concentrated samples or low-solubility analyzes [13].

3.2.6 Concentration of Additives

Adding some modifiers and organic and inorganic additives may increase the solubility of analyzes in the water. Adding some organic solvents such as ethanol to water in some cases leads to higher efficiency access at lower temperatures. The presence of less than 20% ethanol has been shown to increase the water's ability to dissolve many natural substances in plant tissues [15]. Figure 3.1 shows a diagram of the dielectric constant changes of water in terms of temperature with values of 0%, 5% and 20% (v/v) ethanol in aqueous solution [4].

3.3 SOLUBILITY MEASUREMENT METHODS

Laboratory methods of solubility measurement are divided into static and dynamic methods.

3.3.1 A Review of Laboratory Work Performed

3.3.1.1 Static Method

In the static method, a large amount of analyte and a certain volume of solvent are both loaded into an equilibrium chamber, and an equal amount of sample solution is periodically removed from the chamber to balance. Curren and King (2001) used a static method to measure

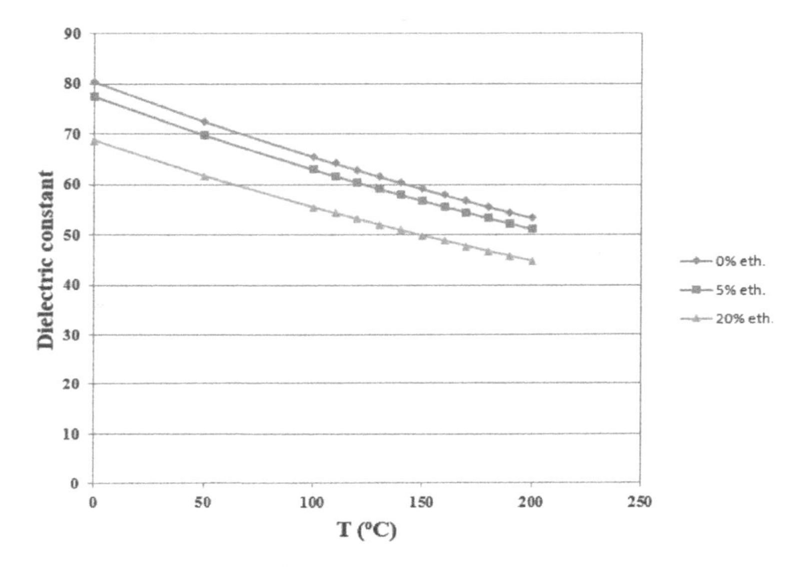

Figure 3.1 Dielectric constant changes of water in terms of temperature and different amounts of ethanol.

(Source: Teoh, W.H., Mammucari, R., and Foster, N. R. (2011). In: *CHEMECA Annual Conference*. 18–21, Sydney, Australia).

the solubility of atrazine, cyanazine and simazine at 50 to 125°C and at a pressure of 50 atm [3]. Figure 3.2 shows the device.

Solubility cell consists of a stainless steel pipe section with an outer diameter of 0.5 inches and a wall thickness of 0.625 inches. All other pipes used in this device are made of stainless steel with an outer diameter of 0.625 inches. Pure or modified water is fed to the solubility cell via an ISCO 100DX injection pump, which is set to maintain a constant pressure. As shown in the figure, the water first passes through a preheater coil in the oven. In addition, a cooling coil after the pump is placed in the line to prevent the solvent from returning to the pump.

The contents of the cell are mixed with a magnetic moving rod (1.3 mm) controlled by a remote stirrer. A six-way valve that is resistant to high pressures up to 3000 psi and temperatures up to 175°C is required for water sampling. When the contents of the cell reach the desired temperature, it is stirred until a balance is reached. The solubility balance changes from 2 to 16 hours with the change in stirring

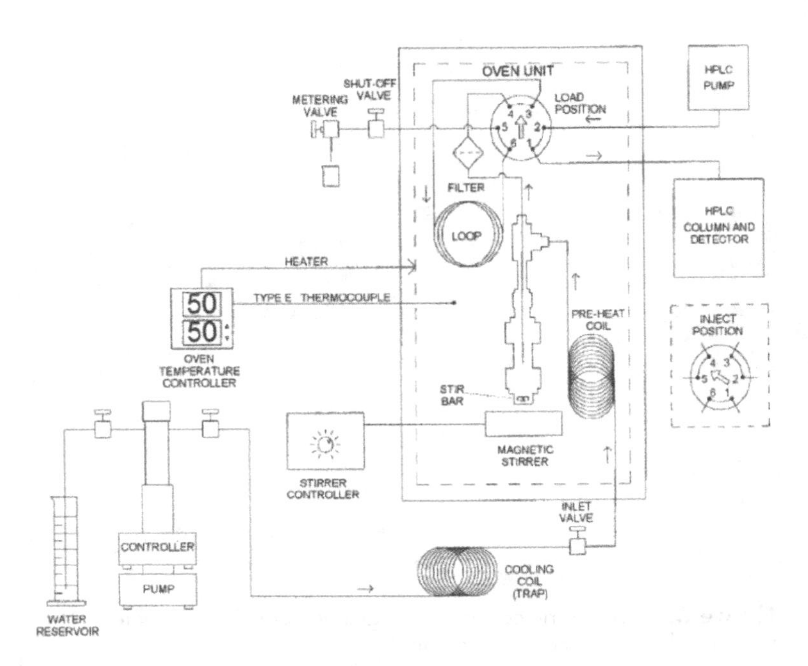

Figure 3.2 The static device used by Curren and King.

(Source: Curren M.S.S. and King, J. W., *Analytical Chemistry* (2001), 73: 740–74).

time until there is not an increase in the measured solubility of the solute at each temperature. Measurements were performed four times at each temperature. Saturated water from the solute is sampled by changing the six-way valve to "load" mode and opening the shut-off valve. A metering valve adjusts the flow in the sample loop. The six-way valve is then turned to "inject" mode and the moving phase directs the sample to the LC system.

The solubility of atrazine in pure and modified water was measured by adding an additional amount of solute to the solutions and stirring them until they reached equilibrium. Each solution was injected into the LC system manually. The following tables show the solubility of atrazine in pure and modified SW at different temperatures and pressures (Tables 3.1 to 3.3).

Table 3.1 shows that with increasing temperature, the dielectric constant of water decreases and the solubility of atrazine increases. Also, according to Tables 3.2 and 3.3, increasing the amount of ethanol

TABLE 3.1
Atrazine Solubility in SW at Different Temperatures and Pressures

Temperature, °C	P, bar	ε	Solubility, μg/mL
50	50	70	1 ± 70
75	50	62	2 ± 210
100	50	55	30 ± 500
125	50	49	70 ± 1780
125	40	49	210 ± 1770

Source: Curren M. S. S. and King, J. W. (2001). *Analytical Chemistry*, 73, 740–745.

TABLE 3.2
Atrazine Solubility in a Mixture of Water and Ethanol at 25°C and 1 Bar Pressure

wt% Ethanol	ε	Solubility, μg/mL
8	74	16 ± 87
25	64	30 ± 450
33	59	60 ± 930
42	54	90 ± 2400

Source: Curren M. S. S. and King, J. W. (2001). *Analytical Chemistry*, 73, 740–745.

TABLE 3.3
Atrazine Solubility in SW with Ethanol at 100°C and 50 Bar Pressure

wt% Ethanol	ε	Solubility, μg/mL
0	55	500 ± 30
8	51	1900 ± 20
12	49	3560 ± 170
16	48	4810 ± 120
20	46	6240 ± 130

Source: Curren M. S. S. and King, J. W. (2001). *Analytical Chemistry*, 73, 740–745.

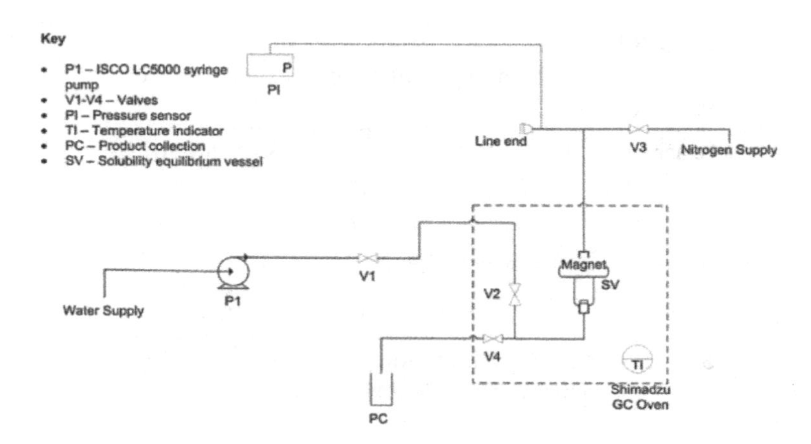

Figure 3.3 Static device for measuring solubility.

(Source: Carr, A. G.; Branch, A.; Mammucari, R. et al. *Journal of Supercritical Fluids* (2010), *55*: 37–42).

reduces the dielectric constant of water and increases the solubility of atrazine. Comparison of Tables 3.2 and 3.3 also shows that the solubility of atrazine in SW compared to water at ambient temperature has increased several times. Foster et al. (2010) used a static method to measure the solubility of budesonide in SW in the temperature range of 25 to 200°C and a pressure of 70 bar (Figure 3.3) [4].

In this study, a batch method was used. Pipes and fittings were stainless steel 316. Solubility tank (SV) has an initial volume of 6.4 mL. Each time it is run, the SV is loaded with an additional soluble amount (about twice the mass of budesonide in a saturated solution at 200°C). Tank is then filled with water by the P1 pump.

"Line end", during the filling period, is kept open and when the water comes out, it is closed with a stainless steel cap. Overflow of water ensures that air escapes from the system. The operating pressure is 70 bar, so that the water remains in the liquid state. The system then reaches the desired temperature using the GC oven. When the desired temperature is reached, the system balances for 10 min with a magnetic stirrer. After stirring for 10 min, the magnetic stirrer is turned off and the nitrogen line is in contact with the solution via V3 (pressure 72 bar).

High nitrogen pressure keeps the tank pressure constant and prevents it from evaporating SW during product collection. The V4 valve opens slightly to allow the SW solution to flow slowly. When nitrogen flows into the collection container, the V4 closes and the oven turns off. The container is then weighed. The V3 valve is closed to remove nitrogen and the pressure of system drops through V4. After the system has cooled, the collection line is removed and washed with 20 mL of acetone to collect the remaining budesonide.

The acetone solution is collected separately in a pre-weighed container. The contents of both containers are dried for 24 h. The water suspension was dried in an oven at 50°C, while the acetone solution was dried in air. The containers were then re-weighed and the amount of water and budesonide extracted from the system was determined. Each experiment was performed three times. The stability of budesonide was investigated during the process at 200°C with FTIR spectroscopy using the KBr disk method. Figure 3.4 shows that the chemical structure of budesonide is maintained during the process. To increase the solubility of budesonide in SW, ethanol and methanol were added to the water. Ethanol in the proportions of 5%, 10% and 20% (v/v) in relation to water and methanol in the ratio of

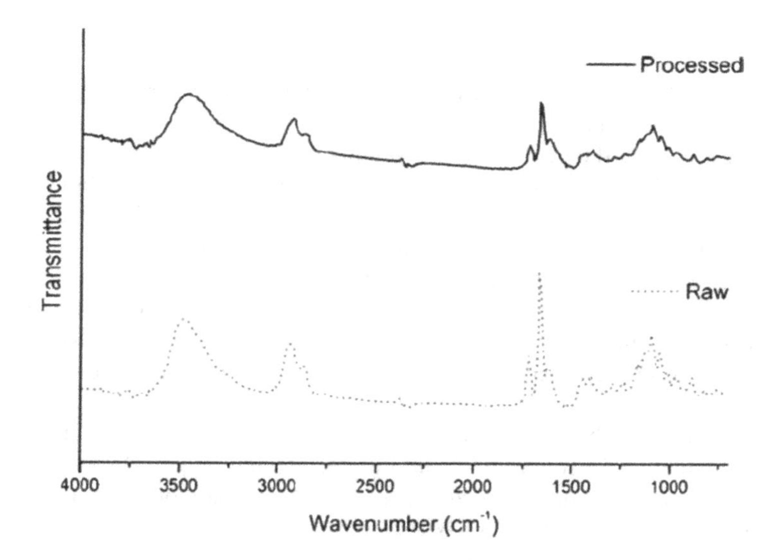

Figure 3.4 FTIR image of budesonide before and after the process in SW at 200°C.

(Source: Carr, A. G.; Branch, A.; Mammucari, R. et al. *Journal of Supercritical Fluids* (2010), 55: 37–42).

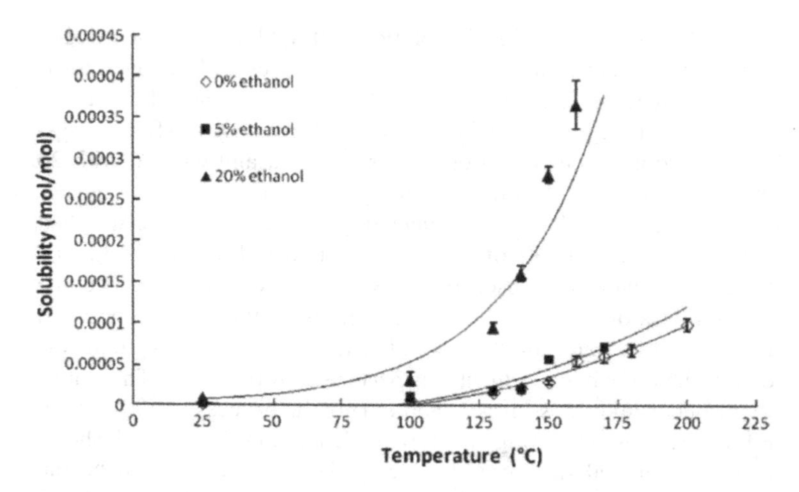

Figure 3.5 Effect of temperature and different amounts of ethanol on the solubility of budesonide.

(Source: Carr, A. G.; Branch, A.; Mammucari, R. et al. *Journal of Supercritical Fluids* (2010), *55*: 37–42).

10%. Dissolved budesonide was measured with UV spectroscopy and with HPLC grade methanol as a solvent.

Figure 3.5 shows the effect of temperature and different amounts of ethanol on the solubility of budesonide. According to this figure, with increasing temperature and increasing the volumetric percentage of ethanol, the solubility of budesonide increases.

In another study, the researchers measured the solubility of anthracene in SW in the temperature range between 393 K and 473 K. In this study, the operating pressure of 50 bar and the nitrogen pressure of 52 bar are reported [16]. The schematic of the device is shown in Figure 3.6.

Figure 3.7 shows the effect of temperature and different amounts of ethanol on the solubility of anthracene, which is observed that with increasing the temperature and increasing the amount of ethanol, the solubility of anthracene in SW increases [17].

Griseofulvin solubility was also measured in SW by Foster et al. (2010) [18]. In 2013, these researchers used a similar method to measure the solubility of anthracene and p-terphenyl in SW in the temperature range of 393 K to 473 K and the pressure range of 50 bar to 150 bar [19]. Also, in 2014, the solubility of multi-ring aromatic

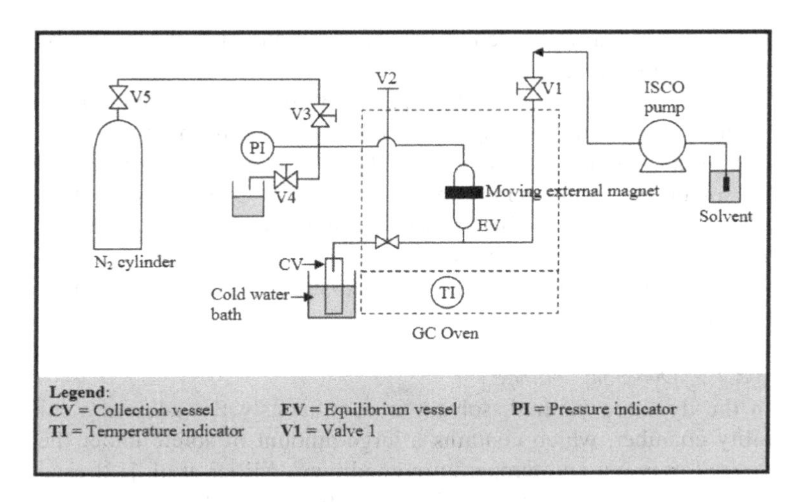

Figure 3.6 The device used by Teoh et al.

(Source: Teoh, W.H., Mammucari, R., and Foster, N. R. (2012). *Ph.D. Thesis*. University of New South Wales).

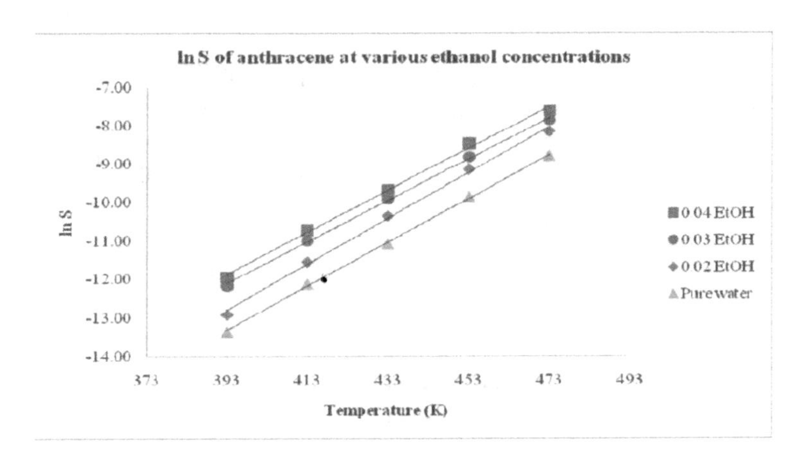

Figure 3.7 The effect of temperature and different amounts of ethanol on the solubility of anthracene.

(Source: Teoh, W.H., Mammucari, R., and Foster, N. R. (2011). In: *CHEMECA Annual Conference*, 18–21, Sydney, Australia).

hydrocarbons was measured in a mixture of water and ethanol at subcritical temperatures [20].

Kayan et al. [21] measured benzoic and salicylic acid solubility in subcritical water using temperature range ranging from 25 to 200°C using static method. Huang et al. [22] used static methods to check the solubility of fatty acids in SW of 80 to 200°C. Kapalavavi et al. [23] using static method investigated the solubility of parabens in SW. With the static method, removing the sample solution from the equilibrium chamber causes a significant pressure drop and disruption of the phase equilibrium process.

3.3.1.2 Dynamic Method

In the dynamic method, solvent is continuously flowed in a solubility chamber, which contains a large amount of solute under the desired pressure and temperature conditions. Miller et al. [12] used the dynamic method, which will follow the shape of the device and the summary of the test method (Figure 3.8). This device was used to measure the solubility of naphthalene, benzopyrene, propazine,

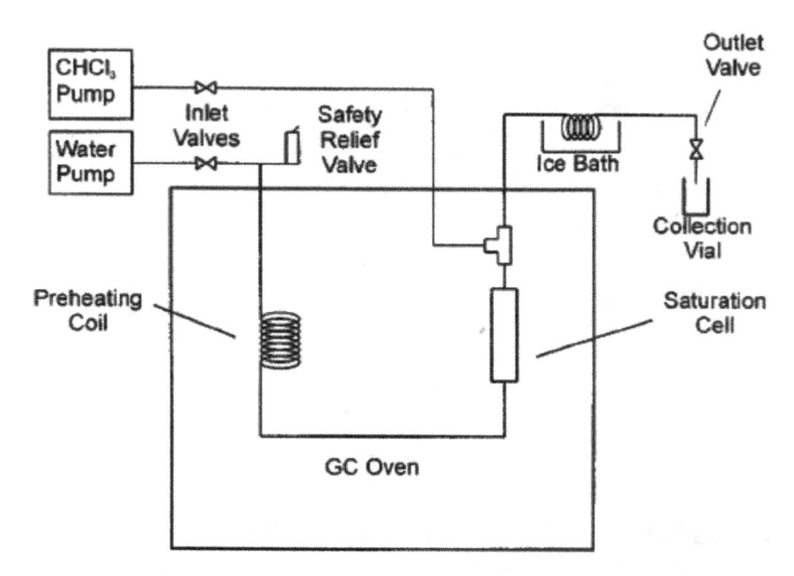

Figure 3.8 Dynamic Solubility Measurement device.

(Source: Miller, D.J. and Hawthorne, S.B. (1998). *Analytical Chemistry,* 70(8), 1618–1621).

chlorthalonil and endosulfan II pesticides in SW in the temperature range between 25 and 250°C.

The GC oven and injection pump (Model 100 D), which operates at a constant flow, is used to provide pressurized water to measure solubility. Water from a preheated tubular steel coil made of stainless steel with dimensions (1/16 in × 0.03 in) 1.6 mm × 762 μm i.d. It enters the saturated tubercle. This tuber contains a sample composition with 10% by weight of pure sand. A preheating coil is essential to ensure that the water has reached operating temperature before entering the saturated cell. The pressure in the saturated tubercle is controlled by the outlet valve. To prevent the solute precipitate in the collection process, an injection pump (Model 260 D) is used to inject chloroform into a three-way lined silica fused "tee" that is in the oven between the saturated tuber and the collecting valve.

A cooling coil made of stainless steel floating in ice water is used to cool the water/chloroform mixture to room temperature. When the water cools in the cooling coil, the saturated organic matter in the water is collected in a 2 mL container. A three-way lined silica fused "tee" and stainless steel pipe with an inner glass coating are used to prevent corrosion of hot water/chloroform. The flow rate of water and chloroform is 0.1 and 0.4 mL/min, respectively. After passing the water through the saturated tuber for a 30 min equilibrium period at each temperature, at least 3 to 5 mL are collected under each condition.

To determine whether water saturation has been achieved in all laboratory conditions, measurements were taken with two different saturation cells with different volumes of 0.498 and 0.83 mL, each with a 10% weight mixture of naphthalene/sand filled. The same results obtained for each tuber indicate that the residence time of the water in the saturated tuber in each of the conditions is sufficient for the water saturation and the solute to be tested. From now on, all measurements were performed using 0.83 mL of cell. Solubility was measured at 25°C and 50°C and then increased to 200 to 250°C to a temperature below the normal melting temperature. Since the melting point of naphthalene is 80°C, the solubility was determined at 25, 35, 50 and 65°C. Pressure of 30–70 bar is used to keep the water in a liquid state at temperatures above 100 °C. This method is also valid by determining the solubility of naphthalene at 25°C and 1 bar. Table 3.4 shows the solubility of naphthalene in water at different temperatures [12].

TABLE 3.4
Naphthalene Solubility in Water at Different Temperatures

Cell Volume (mL)	P bar	Temperature °C	Solubility µg/g
0.498	1	25	36 ± 1
0.498	65	25	33 ± 2
0.83	40	25	34 ± 1
0.83	40	35	49 ± 3
0.83	70	50	101 ± 5
0.83	30	65	216 ± 8

Source: Miller, D.J. and Hawthorne, S.B. (1998). *Analytical Chemistry*, 70(8), 1618–1621.

In general, solubility measurement includes the following steps:

A loaded saturated cell is placed in the GC oven, then the water passes through the preheating coil and reaches a balance at 25°C, and is pumped to the saturated cell with a constant flow rate. Chloroform flow begins when water is released from the collecting valve. The outlet valve is used to ensure that liquid water is present in the entire tested temperature range (30–70 bar). After an equilibrium period (30 min), a minimum of five 3-min fractions are collected at each condition, and the samples are analyzed with GC/FID or GC/MS.

Table 3.5 shows the solubility of propazine, chlorthalonil and endosulfan II pesticides in SW at different temperatures and constant pressures [12].

The experiments were performed at pressures 60–70 bar for propazine and chlorthalonil and 30–40 bar for endosulfan II. As can be seen, the solubility of the tested compounds also increases with increasing temperature. The researchers also used a similar method to measure the solubility of some polycyclic aromatic hydrocarbons, such as anthracene, pyrene, chrysanthemum, perilene and carbazole, in SW [10]. The temperature range tested was 298–498 K and the pressure range was 30–60 bar. The results of the experiments showed that the pressure had little effect, but increasing the temperature to 498 K increased the solubility of the desired compounds up to five times.

King et al. (2012) measured the temperature dependence of the solubility of xylose, glucose and maltose by dynamic method [24].

TABLE 3.5
Solubility of Compounds in Water at Different Temperatures and Constant Pressure

	Solubility, µg/g		
T, °C	Propazine	Chlorothalonil	Endosulfan II
25	6.3 ± 0.5	0.180 ± .01	0.27 ± 0.07
50	13.7 ± 0.8	0.80 ± 0.06	1.1 ± 0.1
100	106 ± 3	28 ± 2	30 ± 3
150	25601 ± 10	950 ± 23	720 ± 96
200	26800 ± 3600	23400 ± 720	4500 ± 530

Source: Miller, D.J. and Hawthorne, S.B. (1998). *Analytical Chemistry*, 70(8), 1618–1621.

The shape of the device used is shown in Figure 3.9. This device is a modified system used by Miller et al. and the system has two high-precision injection pumps to create a constant flow rate of water. Measurement of solubility at constant temperature was performed using an oven (Mode 15890). The water from the storage tank by the

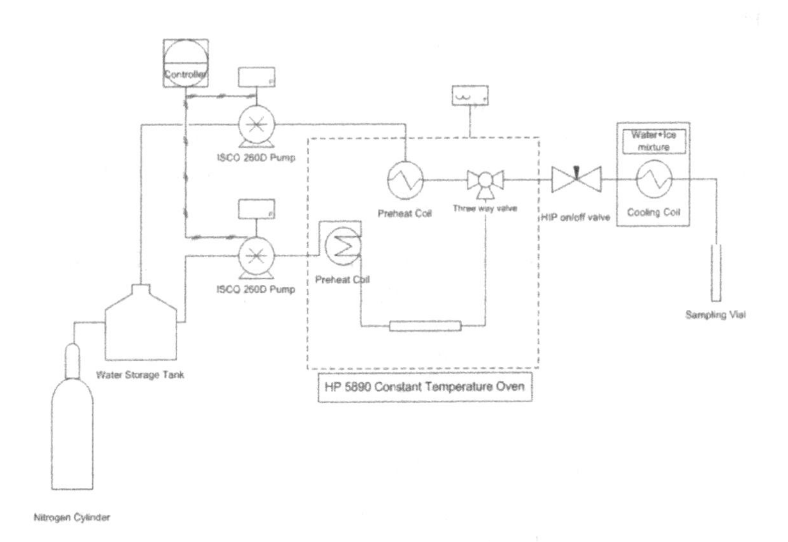

Figure 3.9 Dynamic device used by Srinivas et al.

(Source: Srinivas, K., King, J.W., Howard, L.R. and et al. (2010). *Journal of Food Engineering*, 100(2), 208–218).

injection pump first passes through a preheating coil and then comes in contact with an additional amount of the desired compound in the saturated cell. This saturated tuber is inside the oven. In addition, the water-soluble compound (at a certain laboratory temperature) comes into contact with excess water through a second pump with a constant flow rate. The two streams are mixed in a "T" inside the oven. The flow rates in both injection pumps are controlled by an SFX 200 controller. The solvent flow rate from pump 1 varies depending on the test temperature, and the flow rate in the second pump is set to maintain a dilution factor of 4. Excess water is used to ensure that when the mixture is drained from the oven to room temperature, the compound remains in solution, thus preventing the deposition of the dissolved compound in the pipes.

After exiting the oven, the mixture flows through a cooling coil with an on/off valve and enters the collection container to determine the amount of soluble concentration using high-efficiency liquid chromatography (HPLC). The concentration of compounds in water (g/L) at a given laboratory temperature increases as a function of time until it reaches a saturated value, which is determined by the solubility of the compound in water at the specified laboratory temperature. These researchers used the same device in 2010 to measure the solubility of quercetin and quercetin in water at temperatures 25–140°C [25].

According to Figure 3.10, the solubility of quercetin and quercetin increases with increasing temperature. King et al. [26] also measured the solubility of gallic acid, catechins and protocatechic acid in SW using a dynamic method. In 2015, Yabalak et al. [27] used dynamic methods to measure the solubility of sebacic acid in SW. Zhang et al. [28] used a dynamic method to investigate the solubility of carbohydrates in SW. Takebayashi et al. [29] used a dynamic method to investigate the solubility of terfetal acid in SW at 75–275°C. Karasek et al. measured the solubility of various compounds in SW in a dynamic method in several articles [30–33].

3.4 AN OVERVIEW OF THE MODELING PERFORMED

Solubility data provides us important information for optimizing an extraction process. Solubility forecasting models must accurately predict how matter will dissolve in SW and determine the possibility of a process before testing. The following models are mentioned below.

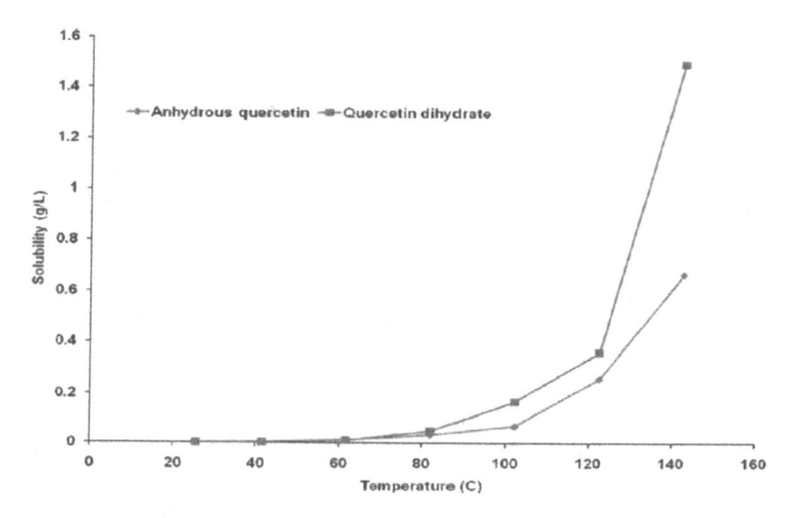

Figure 3.10 Solubility of quercetin and quercetin as temperature functions.

(Source: Srinivas, K., King, J.W., Howard, L.R. and et al. (2010). *Journal of Food Engineering*, 100(2), 208–218).

3.4.1 Experimental and Semi-Experimental Models

This section will focus on models that use solubility data to structure first-, second- or third-degree general relationships, based on room temperature. It uses an optimized model to predict solubility of a SW compound.

Miller et al. (1998) proposed experimental models to predict solubility as a function of temperature [10]. This method only requires knowing the solubility at ambient temperature.

Zeroth approximation:

$$\ln x_2(T) \approx (T_0/T)\ln x_2(T_0) \tag{3.1}$$

In the above equation, $x_2(T)$ is the solubility in the form of molar fraction at temperature T, and $x_2(T_0)$ is solubility at temperature T_0, which is often the ambient temperature. In the above equation, it is assumed that the solute does not absorb water under equilibrium conditions.

Miller changed Eq. (3.1) to Eq. (3.2) in order to match a grade 3 relationship to explain the solubility data for anthracene, pyrene,

TABLE 3.6
Modeling Errors of HOCs Solubility in SW

Temperature	Benzopyrene	Propazine	Chlorthalonil	Endosulfan ii
50°C	0%	–43%	7%	–2%
100°C	0%	–39%	67%	55%
150°C	–26%	–46%	70%	75%
200°C	–33.8	–5%	70%	7%

Source: Miller, D.J., Hawthorne, S.B., Gizir, A.M. et al. (1998). *Journal of Chemical and Engineering Data*, 43(6), 1043–1047. *(Errors Calculated Using Eq. (3.3))*

chrysanthemum, perilene, and carbazol. Eq. (3.2) was then used to predict the solubility of benzopyrene, propazine, chlorthalonil and endosulfan II. The value of the errors is shown in Table 3.6. The error is calculated using Eq. (3.3) and x is the solubility and E is the error value in percent.

First approximation:

$$\ln x_2(T) = (T_0/T)\ln\left[x_2(T_0)\right] + 15\,(T/T_0 - 1)^3 \tag{3.2}$$

These equations are based solely on knowing the solubility at ambient temperature, and the molecular properties of the solute have not been used. Comparing the results of Eq. (3.1) and (3.2) shows that, in general, Eq. (3.1) is slightly better for predicting the solubility of soluble matter in the temperature range between ambient temperature and 373 K, and Eq. (3.2) provides better predictions at temperatures above 373 K.

$$E(\%) = \frac{x(experimental) - x(calculated)}{x(experimental)} \times 100 \tag{3.3}$$

The Miller model predicts solubility well at low temperatures, but cannot predict it correctly at higher temperatures.

King et al. [25] presented the quaternary solubility without water and two waters at different temperatures using the corrected Apelblat relationship [34, 35] as follows:

$$\ln(x_s) = A + \frac{B}{T} + C\ln(T) \tag{3.4}$$

where x_s is the solubility of the compound in water at temperature T (K) as a molar fraction and A, B and C are experimental constants. Del Valle et al. [36] presented a semi-experimental relationship for the solubility of compounds in SW as a function of temperature. They proved the effect of critical temperature and decentralized coefficient of pure solute and obtained good results for different compounds.

Karasek et al. (2006) [37] developed a semi-experimental model for calculating PAHs' solubility in SW. This model was used to calculate the solubility of some PAHs, such as anthracene and naphthalene at 25–250°C, which used only pure water and solute properties. This model shows an average error of 44%. In 2014, Foster et al. [22] presented an experimental model for determining the solubility of anthracene and p-terphenyl in SW. The mentioned research is based on the use of the physical properties of soluble matter and water, and does not consider intermolecular interactions. Therefore, these models are correlative models and not predictive models.

3.4.2 Dielectric Constant Model

Recently, articles on the solubility of organic compounds in SW have been modeled as a direct function of solvent dielectric constant. The dielectric constant of an organic solvent mixture is presented using published data and Akerlof relationships [38]. Akerlof et al. expressed a mathematical relationship for quantifying the effect of ethanol and temperature on the dielectric constant of subcritical ethanol-water systems [38]. This model is not able to predict without at least three known solubility data. If the solubility of the substance at different temperatures in pure water is known, the parameters may be easily adjusted and then used to predict the solubility in a mixture of different solvents at temperatures below 200°C.

Foster et al. [4] modeled the solubility of organic compounds in SW as a direct function of the solvent dielectric constant. By changing the temperature and proportion of organic solvents in the water mixture (with ethanol/methanol in this case), in this model the logarithm solubility of budesonide in water, according to Figure 3.11, was a linear function of the solvent dielectric constant. One of the results of the dependence of solubility on dielectric constant is that the general solubility relationship can be written as the following equation:

$$\ln x_2 = A\varepsilon + B \tag{3.5}$$

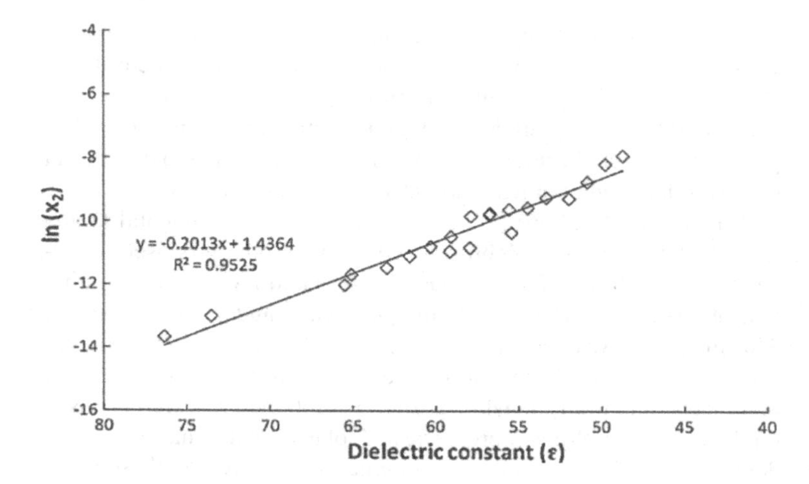

Figure 3.11 The solubility of budesonide in subcritical water solution and modified water with ethanol.

(Source: Carr, A. G.; Branch, A.; Mammucari, R. et al. *Journal of Supercritical Fluids* (2010), *55*: 37–42).

That ε can be calculated from the Akerlof relation using the temperature and mole fraction of ethanol in the solvent. *A* and *B* are linear constants that are regulated by linear regression on constant dielectric-solubility predictions. For budesonide, $A = -0.2013$ and $B = 1.4364$. Using the following equation, the mean error of the dielectric model for budesonide from 25 to 200°C in a water/methanol solution with 0%, 5% and 20% methanol ratios was 0.14%.

$$\text{Error}(\%) = \left| \frac{\ln x_{\text{actual}} - \ln x_{\text{model}}}{\ln x_{\text{actual}}} \right| \times 100 \tag{3.6}$$

3.4.3 Equations of State (EOS)

Thermodynamic models such as state equations can be used to evaluate the solubility of PAHs. These state equations must consider the effects of hydrogen bonding. The most successful state equation of this type is the Cubic Plus Association (CPA) equation [39–40]. The CPA equation works very well for aquatic systems [40–47], and is mathematically simple. CPA EOS can also be successfully used as a tool to predict the solubility of multi-ring aromatic hydrocarbons [48].

CPA EOS is introduced as the following equation:

$$Z = Z^{\text{SRK}} + Z^{\text{association}} \tag{3.7}$$

where Z is compressibility factor. The equation of CPA state and the parameters of this equation will be fully explained in mentioned reference [49].

The thermodynamics of solid-liquid equilibrium and the solubility of S is calculated from the following general statements:

$$\ln \frac{f_s^{\text{liq}}(T,P)}{f_s^{\text{sol}}(T,P)} = \frac{\Delta_{\text{fus}} H}{R}\left(\frac{1}{T} - \frac{1}{T_m}\right) - \frac{\Delta C_p}{R}\left[\frac{T_m}{T} - \ln\left(\frac{T_m}{T}\right) - 1\right] \tag{3.8}$$

where ΔH_{fus} is the solute melting enthalpy, T is the temperature, T_m is the solute melting temperature, ΔC_p is the difference between molar thermal capacities of solid and liquid and R is the global constant of gases. For solid-liquid equilibrium, pressure changes usually do not have a significant effect on equilibrium unless the pressure changes are too large (10–100 MPa). The heat capacity part can be neglected in contrast to the enthalpy part, and using the equation of state, the following sentence can be used for solubility:

$$x_s = \frac{\varphi_s^{Lo}}{\varphi_s^{L}} \exp\left[-\frac{\Delta_{\text{fus}} H}{R}\left(\frac{1}{T} - \frac{1}{T_m}\right)\right] \tag{3.9}$$

φ is the fugitive coefficient and "o" are related to pure matter. From the equation of state, φ is obtained and then x_s or solubility is obtained. Using the CPA equation, Oliveria et al. [50] obtained the solubility of some PAHs in SW and compared them with experimental data from other articles. Their results showed that the CPA equation could be used as a successful tool for predicting the solubility of PAHs in SW. Foster et al. (2013) [19] used from Peng Robinson state equation to determine the solubility of some PAHs in SW.

3.4.4 Regular Solution Theory (RST)

Thermodynamic Framework:

The condition for the phase equilibrium of the solute (2) between the two solid and liquid phases is:

$$f_2^S = f_2^L \tag{3.10}$$

f_2^S is the solute fugacity in the solid phase and f_2^L is the solute fugacity in the liquid phase. If the solid phase is a pure substance, assuming complete solubility of the solvent in the solid phase, the pure solute fugacity in the solid phase, f_2^S, is equal to its fugacity in the solvent.

The solute fugacity in liquid phase is equal to:

$$f_2^L = \gamma_2 x_2 f_2^0 \tag{3.11}$$

In the above equation, γ_2 is the active coefficient of solubility in the liquid phase and x_2 is the molar fraction of solubility and f_2^0 is the standard state fugitive of liquid phase, which is usually the pure liquid fugitive at system temperature and pure liquid vapor pressure. By placing the Eq. (3.11) in the Eq. (3.10) we have:

$$\ln x_2 = \ln\left(f_2^S / f_2^0\right) - \ln\gamma_2 \tag{3.12}$$

There is often a slight difference between the triple point temperature and the normal melting temperature, so the ratio $(f_2^S)/(f_2^0)$ can be written as follows:

$$\ln\left(f_2^s / f_2^0\right) = -\frac{\Delta H_{m2}}{RT_{m2}}\left(\frac{T_{m2}}{T} - 1\right) + \frac{\Delta C_{p2}}{R}\left[\left(\frac{T_{m2}}{T} - 1\right) + \ln\left(\frac{T}{T_{m2}}\right)\right] - \int_{P_{sat}}^{P} \frac{V_2^{liq} - V_2^s}{RT} dP \tag{3.13}$$

T_{m2} and ΔH_{m2} are the normal melting temperature and solute melting enthalpy, respectively, and ΔC_{p2} is the difference between the heat capacity of the pure liquid and the solute solid. The first sentence on the right of the Eq. (3.13) is important. The second sentence is often small, especially when the temperature of the mixture and the temperature of the solute melting point do not differ much. The last sentence of Eq. (3.13), which shows the effect of pressure on solute fugacity, is important only at very high pressures because the difference between the solute molar volume in liquid and solid phases can be ignored. By applying the above standard simplifications, the Eq. (3.13) becomes the following equation:

$$\ln\left(f_2^s / f_2^0\right) = -\frac{\Delta H_{m2}}{RT_{m2}}\left(\frac{T_{m2}}{T} - 1\right) \tag{3.14}$$

And if it is assumed that the mixture of water-solute is ideal ($\gamma_2 = 1$), we will have the following statement for the solubility of ideal:

$$\ln x_2^{id} = -\frac{\Delta H_{m2}}{RT_{m2}}\left(\frac{T_{m2}}{T} - 1\right) \tag{3.15}$$

As a result, the ideal solubility depends only on the melting properties of the solute and the temperature of the system. We have from Eq. (3.12) and Eq. (3.15):

$$\ln x_2 = \ln x_2^{id} - \ln \gamma_2 \tag{3.16}$$

In this theory, the activity coefficient is obtained from the solubility parameters.

According to the Scatchard-Hildebrand regular solution model, the solubility of a solid solute in a liquid is expressed by the following equation [51]:

$$\ln \gamma_2 = \frac{V_2}{RT} \varphi_1^2 (\delta_1 - \delta_2)^2 \tag{3.17}$$

V_2 is the solute molar volume in the liquid state (usually the solid solute melting temperature is higher than the mixed temperature), V_1 is the solute molar volume of the liquid solvent (water) and φ_1 is the volume fraction of water and δ_1 and δ_2 are the solute and water solubility parameters, respectively and they are calculated with the following equations:

$$\varphi_1 = \frac{(x_1 V_1)}{(x_1 V_1 + x_2 V_2)} \tag{3.18}$$

$$\delta_i = \left[\frac{\Delta U_i}{V_i}\right]^{1/2} \qquad I = 1,2 \tag{3.19}$$

ΔU_i is the energy of evaporation and V_i is the liquid molar volume. The solubility parameter may be obtained from the method provided by Fedors [52] and the Hansen solubility parameter [53].

Due to the non-ideal intensity of the PAHs water mixture (which is not only due to the polarity of the water but also due to the asymmetry in the size of the molecules), regular solution theory (RST) gives poor results.

Hansen's solubility parameters (HSP) are used to initiate solubility for polar and non-polar compounds in a variety of solvents. Although this method does not predict solubility, it is able to determine whether the solvent is good or bad (determining whether the solvent is good or not is slightly done) [53]. The HSP method focuses on the partial participation of hydrogen and polar bonding forces and the state of dispersion in the solution of an organic compound in the solvent. Dispersion forces are calculated from a set of contribution factors for organic molecules [53, 54]. Polar forces are calculated using bipolar motion data. Hydrogen bonding forces are considered for the use of evaporative heat data. The participation of these three forces places the solute and the solvent in a three-dimensional space.

Srinivas et al. used HSP to predict the ability of SW to solve a number of complex organic compounds such as glucose and catechins [54]. The inability of this method to calculate an actual amount of solubility limits the use of the HSP model.

3.4.5 UNIFAC-Based Models

This model is based on group-contribution methods. This means that in this method, each molecule is considered as a whole of the subgroups that make up that molecule. Therefore, to calculate the thermodynamic properties of the solution, only the subgroups of the molecules in the solution are considered.

The UNIFAC model was introduced in 1975 by Fredenslund et al. [55] with modifications to the UNIQUAC model. In fact, UNIFAC stands for UNIversal Functional group Activity Coefficient. Because this model is based on the group-contribution method, the non-ideals of complex molecular compounds are expressed on the basis of smaller groups consisting of atoms in the molecules.

In this model, the deviation from the ideal state is proposed for two reasons [56]: (1) Non-ideal due to differences in the size and shape of existing groups, which is called the combinatorial part, and (2) non-ideal resulting from interactions between groups, which they call the residual part.

Therefore, in the UNIFAC model, the activity coefficient is expressed as the sum of the combinatorial and residual parts:

$$\ln \gamma_i^{\text{UNIFAC}} = \ln \gamma_i^C + \ln \gamma_i^R \tag{3.20}$$

The C and R headings represent the combinatorial and residual parts, respectively. The combinatorial part in Eq. (3.20) is as follows:

$$\ln \gamma_i^C = \ln \frac{\varphi_i}{x_i} + 1 - \frac{\varphi_i}{x_i} - \frac{z}{2} q_i \left(\ln \frac{\varphi_i}{\theta_i} + 1 - \frac{\varphi_i}{\theta_i} \right) \tag{3.21}$$

In the above equation, x_i is the molar fraction of part i, θ_i and φ_i are the surface fraction and segment fraction I respectively, which are obtained using the following equations:

$$\varphi_i = \frac{r_i x_i}{\sum_{i=1}^{n} r_i x_i} \tag{3.22}$$

$$\theta_i = \frac{q_i x_i}{\sum_{i=1}^{n} q_i x_i} \tag{3.23}$$

The r_i and q_i are the parameters of the hard core volume (Van der Waals volume) and the surface area (Van der Waals area) of the molecule, respectively, which are calculated as follows:

$$r_i = \sum_{k=1}^{n} \vartheta_k^i R_k \tag{3.24}$$

$$q_i = \sum_{k=1}^{n} \vartheta_k^i Q_k \tag{3.25}$$

ϑ_k^i is the number of groups of type k in component i, and R_k and Q_k are related to the volume and area of group k, which are calculated by Bond relations [57].

Ebrams and Prausnitz considered each polyethylene segment (a unit of CH_2) as the standard segment and presented the relationship between the volume and surface area of the Van der Waals with the parameters R_k and Q_k as follows:

$$R_k = \frac{V_{wk}}{15.17} \tag{3.26}$$

$$Q_k = \frac{A_{wk}}{(2.5\times10^9)} \tag{3.27}$$

V_{wk} and A_{wk} are the Van der Waals group volumes and Van der Waals group surface areas. As can be seen in the above equations, $\ln\gamma_i^C$ does not depend on temperature. For the combinatorial and residual part, the activity coefficient of the following equation is used:

$$\ln\gamma_i^R = \sum_k \vartheta_k^i\left(\ln\Gamma_k - \ln\Gamma_k^i\right) \tag{3.28}$$

Γ_k is the residual activity coefficient and Γ_k^i is the combinatorial activity coefficient of group k in the reference solution that contains only type i molecules. The residual activity coefficient of group k is obtained like the residual part of the UNIQUAC equation:

$$\ln\Gamma_k = Q_k\left[1-\ln\left(\sum_m \Theta_m \Psi_{mk}\right) - \sum_m \frac{\Theta_m \Psi_{km}}{\Sigma_n \Theta_m \Psi_{nm}}\right] \tag{3.29}$$

$k,\ m,\ n = 1,2,\ldots N$ (all groups)

In the above relation, Θ_m is the surface fraction of the group m, which is calculated as follows:

$$\Theta_m = \frac{Q_m X_m}{\Sigma_n Q_n X_n} \tag{3.30}$$

X_m is the molar fraction of group m in the mixture:

$$X_m = \frac{\Sigma_j \vartheta_m^j X_j}{\Sigma_j \Sigma_n \vartheta_n^j X_j} \quad (3.31)$$

Ψ_{mn} is the interaction parameter is the group that is calculated by the following equation:

$$\Psi_{mn} = \exp\left[-\frac{U_{mn} - U_{nm}}{RT}\right] = \exp\left(-\frac{a_{mn}}{T}\right) \quad (3.32)$$

U_{mn} is the amount of energy interacting between groups m and n. The group interaction parameter, a_{mn}, should be obtained from the laboratory phase equilibrium data. a_{mn} has the Kelvin dimension and $a_{nm} \neq a_{mn}$.

Fornari et al. [58] examined the ability of UNIFAC models to predict the solubility of PAHs in SW as a function of temperature. The original UNIFAC [54], its modified model (Dortmund) [59] and model A-UNIFAC (associative) [60] are used in the temperature range 298–500 K. These models are based on the group contribution that consider intermolecular energy interactions. The best predictions of solubility data were obtained using the modified UNIFAC (Dortmund) model. In addition, the application of Model A-UNIFAC confirms this hypothesis that the reduction of cumulative effects between SW molecules (according to decrease in dielectric constant with temperature, i.e., a decrease in water polarity) improves solubility of non-polar hydrophobic organic compounds to a large extent and this is what is confirmed by empirical observations. In fact, the A-UNIFAC model, which considers the cumulative interactions (hydrogen bonding) between water molecules, is used as a predictor method, but in general, the results of Model A-UNIFAC are not correctly the Model M-UNIFAC. Hansen et al. [61] also emphasized the good predictions of this model for estimating the solubility of anthracene, fluorescent and pyrene in polar solvents.

Fornari et al. [62] also compared the ability of two thermodynamic models, the regular solution theory and UNIFAC-based models, to predict the solubility of PAHs in SW. Specifically, the modified UNIFAC (Dortmund) model demonstrated good relation between temperature and solubility. Foster et al. [63] correlated the naproxen solubility in SW with temperature using the modified activity model (M-UNIFAC) and model errors were optimized by optimizing the water-carboxylic acid interaction parameter. In 2014, the researchers also compared UNIFAC models to predict the solubility of some PAHs in SW [22].

Over the past ten years, models have become more precise with the increase in solubility databases. As more data becomes available, models will be applicable to a wider range of organic compounds. Therefore, with increasing solubility data for compounds with similar structure and diverse groups, the M-UNIFAC model may be developed in later stages.

REFERENCES

1. Terhi, A. (2007) Parameters affecting the extraction of polycyclic aromatic hydrocarbons with pressurised hot water, Academic Dissertation, University of Helsinki, Department of Chemistry, Laboratory of Analytical Chemistry, Finland.
2. Kubátová, A., Jansen, B., Vaudoisot, J.F., Hawthorne, S.B. (2002) Thermodynamic and kinetic models for the extraction of essential oil from savory and polycyclic aromatic hydrocarbons from soil with hot (subcritical) water and supercritical CO_2, *Journal of Chromatography A*, 975(1), 175–188.
3. Curren M. S. S., and King, J. W. (2001) Solubility of Triazine Pesticides in Pure and Modified Subcritical Water, *Analytical Chemistry*, 73, 740–745.
4. Carr, A. G., Branch, A., Mammucari, R., Foster, N. R. (2010) The solubility and solubility modelling of budesonide in pure and modified subcritical water solutions, *Journal of Supercritical Fluids*, 55, 37–42.
5. Tangkhavanich, B., Kobayashi, T., Adachi, Sh. (2014). Effects of repeated treatment on the properties of rice stem extract using subcritical water, ethanol, and their mixture, *Journal of Industrial and Engineering Chemistry*, 20, 2610–2614.
6. Franks, F. (1983) *Water*, 1st ed., The Royal Society of Chemistry, London.
7. Nakahara, M., Matubayasi, N., Wakai, C., Tsujino, Y. (2001) Structure and dynamics of water: from ambient to supercritical, *Journal of Molecular Liquids*, 90(1–3), 75–83.
8. Caffarena, E.R., Grigera, J.R. (2004) On the hydrogen bond structure of water at different densities, *Physica A*, 342(1–2), 34–39.
9. Shinoda, K. (1977) Iceberg formation and solubility, *The Journal of Physical Chemistry*, 81(13), 1300–1302.

10. Miller, D.J., Hawthorne, S.B., Gizir, A.M., Clifford, A.A. (1998) Solubility of polycyclic aromatic hydrocarbons in subcritical water from 298 K to 498 K, *Journal of Chemical and Engineering Data*, 43(6), 1043–1047.
11. Khuwijitjaru, P., Adachi, S., Matsuno, R. (2002) Solubility of saturated fatty acids in water at elevated temperatures, *Bioscience, Biotechnology and Biochemistry*, 66(8), 1723–1726.
12. Miller, D.J., Hawthorne, S.B. (1998) Method for determining the solubilities of hydrophobic organics in subcritical water, *Analytical Chemistry*, 70(8), 1618–1621.
13. Carr A. G., Mammucari R., Foster N.R., 2011. A review of subcritical water as a solvent and its utilisation for the processing of hydrophobic organic compounds, *Chemical Engineering Journal*, 172, 1–17.
14. Juhani, K., Hartonen, K., and Riekkola, M. (2007) Analytical extractions with water at elevated temperatures and pressures, *Trends in Analytical Chemistry*, 26(5), 396–412.
15. H.S. Mohammadi, A. Haghighi Asl, M. Khajenoori (2020). Experimental measurement and correlation of solubility of β-carotene in pure and ethanol-modified subcritical water. *Chinese Journal of Chemical Engineering*, 28 (10), 2620–2625.
16. Teoh, W.H., Mammucari, R., and Foster, N. R. (2012). Fundamental solubility study of polycyclic aromatic hydrocarbons in subcritical water and ethanol mixtures. *Ph.D. Thesis*. University of New South Wales.
17. Teoh, W.H., Mammucari, R., and Foster, N. R., 2011. Ternary solubilities of anthracene in pressurized binary water/ethanol mixtures at temperatures ranging from 393 K to 473 K. *CHEMECA Annual Conference* (2011) 18–21, Sydney, Australia.
18. Carr, A. G., Mammucari, R., and Foster, N. R. (2010) Solubility and micronization of griseofulvin in subcritical water, *Industrial & Engineering Chemistry Research*, 49, 3403–3410.
19. Teoh, W.H., Mammucari, R., Vieira de Melo, S.A.B., Foster, N.R. (2013) Solubility and solubility modeling of polycyclic aromatic hydrocarbons in subcritical water, *Industrial & Engineering Chemistry Research*, 52, 5806–5814.
20. Teoh, W.H., Vieira de Melo, S.A.B., Mammucari, R., Foster, N. R. (2014) Solubility and solubility modeling of polycyclic aromatic hydrocarbons in subcritical ethanol and water, mixtures, *Industrial & Engineering Chemistry Research*, 53, 10238–10248.
21. Kayan, B., Yang, Y., Lindquist, E.J., Gizir, A.M. (2009) Solubility of benzoic and salicylic acids in subcritical water at temperatures ranging from (298 to 473) K., *Journal of Chemical & Engineering Data*, 55(6), 2229–2232.

22. Huang, P.P., Yang, R.F., Qiu, T.Q., Fan, X.D. (2013) Solubility of fatty acids in subcritical water, *The Journal of Supercritical Fluids*, 81, 221–225.
23. Kapalavavi, B., Ankney, J., Baucom, M., Yang, Y. (2014) Solubility of parabens in subcritical water, *Journal of Chemical & Engineering Data*, 59(3), 912–916.
24. King J. W., Srinivas K. (2012) Measurement of aqueous solubility of compounds at high temperature using a dynamic flow apparatus and a Teledyne Isco syringe pump, Syringe Pump Application Note AN27.
25. Srinivas, K., King, J. W., Howard, L. R., Monrad, J. K. (2010) Solubility and solution thermodynamic properties of quercetin and quercetin dihydrate in subcritical water, *Journal of Food Engineering*, 100(2), 208–218.
26. Srinivas, K., King, J. W., Howard, L. R., and Monrad, J. K. (2010) Solubility of gallic acid, catechin and protocatechuic acid in subcritical water from (298.75 to 415.85) K, *Journal of Chemical & Engineering Data*, 55, 3101–3108.
27. Yabalak, E., Gormez, O., Gozmen, B., Gizir, A. M. (2015) The solubility of sebacic acid in subcritical water using the response surface methodology, *International Journal of Industrial Chemistry*, 6(1), 23–29.
28. Zhang, D., Montanes, F., Srinivas, K., Fornari, T., Ibanez, E., King, J.W. (2010) Measurement and correlation of the solubility of carbohydrates in subcritical water, *Industrial & Engineering Chemistry Research*, 49(15), 6691–6698.
29. Takebayashi, Y., Sue, K., Yoda, S., Hakuta, Y., Furuya, T. (2012) Solubility of terephthalic acid in subcritical water, *Journal of Chemical & Engineering Data*, 57(6), 1810–1816.
30. Karásek, P., Hohnová, B., Planeta, J., and Roth, M. (2010) Solubility of Solid Ferrocene in Pressurized Hot Water, *Journal of Chemical & Engineering Data*, 55, 2866–2869.
31. Karásek, P., Hohnová, B., Planeta, J., Šťavíková, L., Roth, M. (2013) Solubilities of selected organic electronic materials in pressurized hot water and estimations of aqueous solubilities at 298.15 K, *Chemosphere*, 90, 2035–2040.
32. Karásek, P., Planeta, J., and Roth, M. (2008) Solubilities of Triptycene, 9-Phenylanthracene, 9,10-Dimethylanthracene, and 2-Methylanthracene in Pressurized Hot Water at Temperatures from 313 K to the Melting Point, *Journal of Chemical & Engineering Data*, 53, 160–164.
33. Karásek, P., Planeta, J., and Roth, M. (2008) Solubilities of Adamantane and Diamantane in Pressurized Hot Water, *Journal of Chemical & Engineering Data*, 53, 816–819.
34. Heryanto, R., Hasan, M., Abdullah, E.C., Kumoro, A.C. (2007) Solubility of stearic acid in various organic solvents and its

prediction using non-ideal solution models, *ScienceAsia*, 33, 469–472.
35. Wang, S., Chen, D. (2006) Solubility of piperonal in different pure solvents and binary isopropanol+water solvent mixtures. *Korean Journal of Chemical Engineering*, 23(6), 1034–1036.
36. Valle, J. M., Fuente, J. C., and King, J. W. (2006) Correlation for the variation with temperature of solute solubilities in hot pressurized water, Proceedings of the VII Iberoamerican Conference on Phase Equilibria and Fluid Properties for Process Design (2006) Morelia, Michoacán, Mexico, October 21–25 pp. 142.
37. Karásek, P., Planeta, J., and Roth, M. (2006) Solubility of Solid Polycyclic Aromatic Hydrocarbons in Pressurized Hot Water at Temperatures from 313 K to the Melting Point, *Journal of Chemical & Engineering Data*, 51, 616–622.
38. Akerlof, G. (2002) Dielectric constants of some organic solvent-water mixtures at various temperatures, *Journal of American Chemical Society*, 54(11), 4125–4139.
39. Economou, I. G. (2002). Statistical associating fluid theory: A successful model for the calculation of thermodynamic and phase equilibrium properties of complex fluid mixtures. *Industrial & Engineering Chemistry Research*, 41(5), 953–962.
40. Kontogeorgis, G. M., Michelsen, M. L., Folas, G. K.; Derawi, S., von Solms, N., Stenby, E. H. (2006) Ten years with the CPA (Cubic-Plus-Association) equation of state. Part 1. Pure compounds and self-associating systems. *Industrial & Engineering Chemistry Research*, 45(14), 4855–4868.
41. Kontogeorgis, G. M., Michelsen, M. L., Folas, G. K., Derawi, S., von Solms, N., Stenby, E. H. (2006) Ten years with the CPA (Cubic-Plus-Association) equation of state. Part 2. Cross-associating and multicomponent systems. *Industrial & Engineering Chemistry Research*, 45(14), 4869–4878.
42. Oliveira, M. B., Coutinho, J. A. P., Queimada, A. J. (2007) Mutual solubilities of hydrocarbons and water with the CPA EoS. *Fluid Phase Equilibria*, 258(1), 58–66.
43. Folas, G. K., Kontogeorgis, G. M., Michelsen, M. L., Stenby, E. H. (2006) Application of the cubic-plus-association (CPA) equation of state to complex mixtures with aromatic hydrocarbons, *Industrial & Engineering Chemistry Research*, 45(4), 1527–1538.
44. Kontogeorgis, G. M., Yakoumis, I. V., Meijer, H., Hendriks, E., Moorwood, T. (1999) Multicomponent phase equilibrium calculations for watermethanol-alkane mixtures, *Fluid Phase Equilibria*, 160, 201–209.
45. Voutsas, E. C., Boulougouris, G. C., Economou, I. G., Tassios, D. P., 2000. Water/hydrocarbon phase equilibria using the thermodynamic perturbation theory, *Industrial & Engineering Chemistry Research*, 39(3): 797–804.

46. Derawi, S. O., Michelsen, M. L., Kontogeorgis, G. M., Stenby, E. H. (2003) Application of the CPA equation of state to glycol/hydrocarbons liquid–liquid equilibria, *Fluid Phase Equilibria*, 209(2), 163–184.
47. Derawi, S. O., Kontogeorgis, G. M., Michelsen, M. L., Stenby, E. H. (2003) Extension of the cubic-plus-association equation of state to glycolwater cross-associating systems, *Industrial & Engineering Chemistry Research*, 42(7), 1470–1477.
48. Oliveira, M. B., Varanda, F. R., Marrucho, I. M., Queimada, A. J., Coutinho, J. A. P. (2008) Prediction of water solubility in biodiesel with the CPA Equation of State, *Industrial & Engineering Chemistry Research*, 47(12), 4278–4285.
49. Mottahedin, P., Haghighi Asl, A., Lotfollahi, M.N. (2017) Experimental and modeling investigation on the solubility of β-carotene in pure and ethanol-modified subcritical water, *Journal of Molecular Liquids*, 237, 257–265.
50. Oliveira, M. B., Oliveira, V. L., Coutinho, J. A. P., Queimada, A. J. (2009) Thermodynamic Modeling of the Aqueous Solubility of PAHs, *Industrial & Engineering Chemistry Research*, 48, 5530–5536.
51. Prausnitz J. M. (1999) *Molecular Thermodynamics of Fluid-Phase Equilibria*, 3rd ed.; Prentice-Hall Inc.: Upper Saddle River, NJ.
52. Fedors, R. F. (1974) A method for estimating both the solubility parameters and molar volumes of liquids, *Polymer Engineering & Science*, 14, 147–154.
53. Hansen, C. M. (1969). The Universality of the Solubility Parameter, *Industrial and Engineering Chemistry Product Research and Development*, 8, 2–11.
54. Srinivas, K., King, J.W., Hansen, C.M. (2008) Prediction and Modeling of Solubility Phenomena in Subcritical Fluids Using an Extended Solubility Parameter Approach, *ACS-AIChE National Meeting*, Spring April 6-10, New Orleans, Louisiana.
55. Fredenslund A., Jones R. L., Prausnitz J. M. (1975) Group-contribution estimation of activity coefficients in nonideal liquid mixtures, *AIChE Journal*, 21, 1086–1099.
56. Gonzalez H. E., Abildskov J., Gani G. (2007) A Method for Prediction of UNIFAC Group Interaction Parameters, *AIChE Journal*, 53(6), 1620–1632.
57. Bondi, A. (1968) *Physical properties of molecular crystals, liquid and gases*, Wiley, New York.
58. Fornari T., Stateva R. P., Señorans F. J., Reglero G., Ibañez E. (2008). Applying UNIFAC-based models to predict the solubility of solids in subcritical water, *The Journal of Supercritical Fluids*, 46, 245–251.

59. Gmehling J., Li J., Schiller M. (1993) A modified UNIFAC model. 2. present parameter matrix and results for different thermodynamic properties, *Industrial & Engineering Chemistry Research*, 32, 178–193.
60. Mengarelli A. C., Brignole E. A., Bottini S. B. (1999) Activity coefficients of associating mixtures by group contribution, *Fluid Phase Equilibria*, 163, 195–207.
61. Hansen H. K., Riverol C., Acree Jr. W. E. (2000) Solubilities of anthracene, fluoranthene and pyrene in organic solvents: comparison of calculated values using UNIFAC and modified UNIFAC (Dortmund) models with experimental data and values using the Mobile Order Theory, *Canadian Journal of Chemical Engineering*, 78, 1168–1174.
62. Fornari T., Ibañez E., Reglero G., Stateva R. P. (2011) Analysis of Predictive Thermodynamic Models for Estimation of Polycyclic Aromatic Solid Solubility in Hot Pressurized Water, *The Open Thermodynamics Journal*, 5, 40–47.
63. Carr A.G., Mammucari R., Foster N. R. (2010) Solubility, Solubility Modeling, and Precipitation of Naproxen from Subcritical Water Solutions, *Industrial & Engineering Chemistry Research*, 49(19), 9385–9393.

Chapter **4**

MODELING OF SUBCRITICAL WATER EXTRACTION

4.1 INTRODUCTION

Modeling is essential for optimizing and designing any separation process. Modeling a complex phenomenon such as the extraction of natural materials is an important activity due to its proposed economic capabilities. Such predictions need to build a model that predicts phase behavior, equilibrium, solubility, adsorption, desorption and so on.

The extraction system with subcritical water includes solvent phase and solid phase. The solvent phase is subcritical water and includes the soluble extract and the solid phase is the plant matrix. Most subcritical water extractions (SWEs) are carried out as a continuous flow of solvent in solid-phase substrates. Mass transfer takes place between two phases and materials are transferred from solid to solvent phase.

Chemical compounds in plants can be synthesized inside the cell or outside the cell wall. The process of SWE of solids generally includes the following steps, Figure 4.1 [1]:

1. Rapid entry of fluid into the matrix;
2. Desorption of analytes from active sites;
3. The penetration of analytes into organic matter;
4. Penetration of analytes into static fluid in porous materials;
5. Penetration of analytes into resident layers of particles outside; and
6. Washing the analytes with a flowing fluid.

To model the extraction process of any compound from any plant particle, only the most important steps are considered as follows:

1. Transfer of compound within the plant matrix (internal penetration);

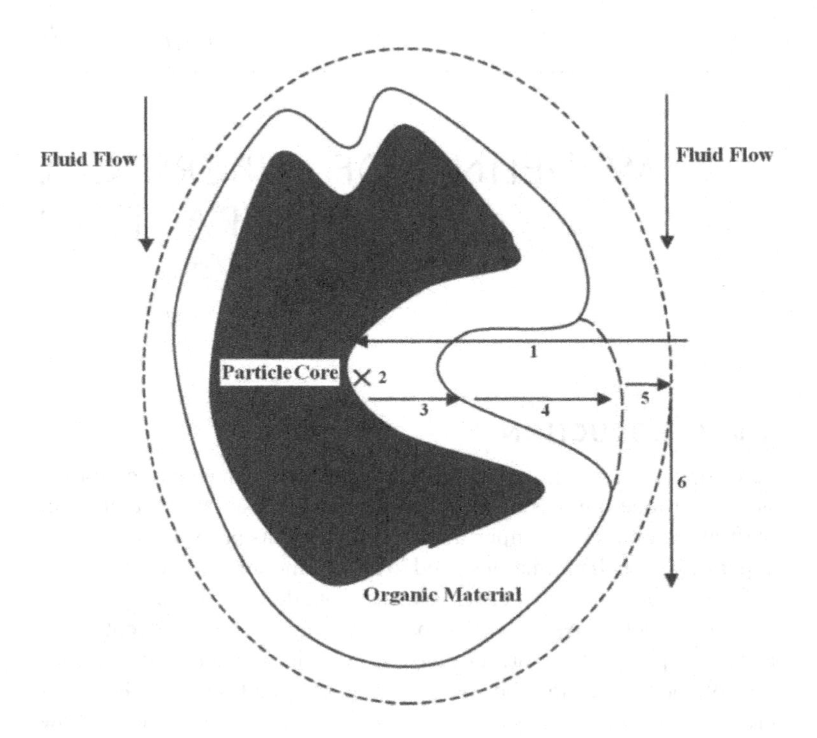

Figure 4.1 Schematic representation of the proposed SWE steps: (1) Rapid entry of fluid into the matrix; (2) desorption of analytes from active sites; (3) the penetration of analytes into organic matter; (4) penetration of analytes into static fluid in porous materials; (5) penetration of analytes into resident layers of particles outside; and (6) washing the analytes with a flowing fluid.

(Source: Haghighi Asl, A. and Khajenoori, M. (2013). *Chapter Book, InTech*, 459–487).

2. Penetration inside the film layer residing around the solid plant particle (external penetration); and
3. Wash the mixture from the plant matrix with thermodynamic separation into the current solvent.

Parameters such as dispersion of the extracted material concentration along the extraction vessel, time, concentration of the extractable material in the bulk phase and radial and axial dispersion, sample preparation and operating parameters are important for the extraction processes.

4.2 INVESTIGATION OF EXISTING MODELS IN SWE

Modeling of SWE is still in its infancy and there are limited articles in this field. From the study of studies, the proposed models so far can be divided as follows:

- Thermodynamic model.
- Synthetic models.
- Single-location kinetic models.
- Two-dimensional kinetic models.
- Thermodynamic separation model with external mass transfer resistance.
- Model based on differential mass balance equations.

4.2.1 Thermodynamic Model

This model is in fact a simple thermodynamic model based on only the K_D thermodynamic distribution coefficient. In this model, it is assumed that the kinetics of the initial desorption step and the separation of the lattice fluid are fast and have no significant effect on the extraction rate compared to leaching. K_D is defined as follows:

$$K_D = \frac{\textit{Concentration of solute in the matrix}}{\textit{Concentration of solute in the extraction fluid}}; \quad \textit{at equilibrium} \tag{4.1}$$

In principle, the mass of the analyte in each unit is the mass of the solvent extracted, and the mass of the analyte remaining in the matrix during the extraction time is determined for the total extraction time, based on the amount of K_D determined for each compound. Therefore, if the K_D model is used for a particular extraction, the shape of the extraction curve is defined as follows [2]:

$$\frac{S_b}{S_o} = \left[\left(1 - \frac{S_a}{S_o}\right) / \left(\frac{K_D m}{(V_b - V_a)\rho} + 1\right)\right] + \frac{S_a}{S_o} \tag{4.2}$$

S_a is the cumulative mass of analyte extracted after volume V_a (mL); S_b is the cumulative mass of the analyte extracted after the volume V_b (information point V_b and S_b information point after V_a and S_a); So, the total mass of the analyte in the plant network, S_b/S_o and S_a/S_o, the cumulative components of the analyte extracted with the extracted solvent after the volume V_b and V_a, respectively; ρ, density of extraction solvent under given conditions (g/mL); *m* is the mass of the extraction sample (g).

It is inferred from the model that the cumulative amount of the extracted solute is described as a function of the volume of solvent passing through the extractor and does not depend on the extraction time. If the solvent flow rate is doubled (other parameters are constant), the amount of passing solvent is doubled for the same unit time, which increases the amount of solute extracted. Therefore, if the cumulative amount of extracted solute is plotted in terms of extraction time, the curve for higher current intensities is higher than lower current intensities. On the other hand, if the same data are plotted by volume, the theoretical curves of all current intensities are completely superimposed.

4.2.2 Kinetic Absorption Model

4.2.2.1 One-Site Kinetic Desorption Model

This model describes extractions that are controlled by the internal penetration of particles. This occurs when the fluid flow is fast enough that the concentration of a particular analyte is well below its thermodynamically controlled range. The one-site kinetic desorption model is based on the mass transfer model, which is similar to the hot ball heat transfer model [3], assuming that the composition distribution is initially uniform in the plant network and the concentration of the compound at the plant network level is zero (there is no solubility limit). For a spherical matrix, the mass ratio of the compound remaining after the extraction time t in the particle (S_r) to the initial mass of the compound (S_o) is given as follows:

$$\frac{S_r}{S_o} = \frac{6}{\pi^2} \sum_{n=1}^{\infty} \frac{1}{n^2} \exp\left(-D_e n^2 \pi^2 t / r^2\right) \tag{4.3}$$

n is the integer, D_e is the effective diffusion coefficient of the compound in the spherical material (m^2/s). The curve tends to be linear for long periods of time (usually after $t > 0.5\ t_c$) and $Ln\ (S_r/S_o)$ is roughly as follows:

$$Ln\,(S_r/S_o) = -0.4977 - t/t_c \tag{4.4}$$

t_c is defined in terms of minutes as follows:

$$t_c = r^2 / \pi^2 D_e \tag{4.5}$$

The extraction model gives y width for the linear curve. In practice, however, the value of width y depends on the shape and size distribution of the particle as well as the soluble distribution within the matrix of solid particles (i.e., the analyte is mainly near the surface or inside the particle) [3]. Another form of Eq. (4.4), also called the one-site kinetic desorption model, can be given as the mass ratio of the removed analyte after time t to the initial mass so:

$$\frac{S_t}{S_o} = 1 - e^{-kt} \tag{4.6}$$

S_t is the mass of the analyte removed by the extraction solvent after time t (mg/g) and S_o is the initial mass of the total analyte in the matrix (mg/g). S_t/S_o is the component of the extracted analyte after time t and k is the first-order velocity constant that describes the extraction.

4.2.2.2 Two-Site Kinetic Desorption Model

This model is a simple deformation of the one-site desorption model. In the previous model, the extraction curve is exponential. At first, the extraction has a steep slope and then it continues with a gentle slope. This may indicate that some analytes separate faster and others more slowly, depending on the degree to which the analytes are connected to the matrix. Therefore, the kinetic desorption model needs two steps to define the extraction curve. A special component of analytes (F) is absorbed at a fast velocity, defined by the velocity constant k_1 (min^{-1}) and the remaining component $(1 - F)$ at a slower velocity, defined by the velocity constant k_2 (min^{-1}). The model is expressed as follows [2]:

$$\frac{S_t}{S_0} = 1 - \left[Fe^{-k_1 t} \right] - \left[(1 - F)e^{-k_2} t \right] \tag{4.7}$$

t is the time (min), S_t is the mass of the analyte removed by the extraction fluid after time t, S_o is the initial mass of the analyte in the matrix. S_t/S_o is the analyte extracted after time t. The kinetic model does not include any factor expressing the current intensity and depends only on time. Therefore, if the extraction efficiency is controlled by the kinetics of the initial desorption step (the rest of the parameters are constant), doubling the flow intensity has little effect on the extraction efficiency.

4.2.3 Thermodynamic Separation with External Mass Transfer Resistance Model

This model describes extraction controlled by external mass transfer. Speed is defined by the resistance type model as follows [3]:

$$\frac{\partial c_s}{\partial t} = -k_e a_p \left[(c_s / K_D) - c\right] \quad (4.8)$$

c is fluid phase concentration (mol/m^3), c_s is solid phase concentration (mol/m^3), k_e is external mass transfer coefficient (m/min), a_p is particle specific surface area (m^2/m^3).

If the analyte concentration in the fluid mass is assumed to be low and the analyte concentration in the liquid at the solid matrix surface is defined by the equilibrium coefficient (K_D), the solution of Eq. (4.8) for the analyte concentration in the solid matrix (c_s) is as follows [3]:

$$c_s = c_o \exp\left(-k_e a_p / K_D \ t\right) \quad (4.9)$$

Due to the difficulty of accurately measuring a_p, a_p and k_e are usually considered as ($a_p k_e$), which is called the volumetric mass transfer coefficient. Factors that affect the amount of $a_p k_e$ include the speed of water flow inside the extractor and the size and shape of the plant sample.

Kubatova et al. (2002) investigated the mechanism controlling the rate of extraction of essential oils from clove and polycyclic hydrocarbons from contaminated soil with two different extraction solvents of subcritical water and supercritical carbon dioxide [2]. They extract each compound from the solid matrix in two steps: (1) Desorption of the compound from its original bonded sites (on) the solid matrix (usually modeled by the diffusion process), and (2) wash the composition of the sample by a method similar to chromatographic elution (controlled by the coefficient of thermodynamic resolution K_D) and from two simple thermodynamics and two-site desorption, to describe the co-profiles used. In both models, data matching method with Excel software is used.

Based on previous discussions, the relative importance of K_D and desorption kinetics were determined by comparing the effects of current intensity changes on the extraction rate of the same samples. Figure 4.2 shows the extraction curves of major essential oil compounds with two different solvents at different flow intensities. They proposed the washing of analytes from the matrix as the predominant mechanism for extraction with subcritical water. Because the increase

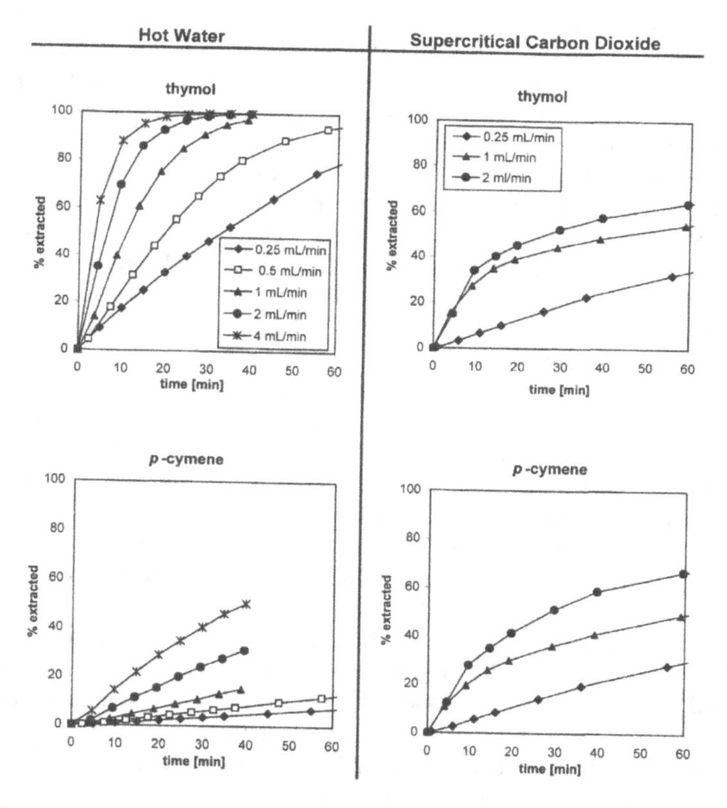

Figure 4.2 Extraction curves of major constituents of clove essential oil with subcritical water and supercritical CO_2 at different current intensities.

(Source: Kubatova, A., Jansen, B., Vaudoisot, J.F. et al. (2002). *Journal of Chromatography A.*, 975 (1): 175–188).

in extraction rate was proportional to the increase in water flow intensity, the temperature was subcritical, regardless of whether the compound was extracted rapidly (i.e., thymol) or slow (i.e., p-cymene). This was also confirmed by simple deletion calculations based on K_D only (for each essential oil component) which fits well with the experimental data for flow intensities of 0.25–4 mL/min.

In contrast, supercritical CO_2 extraction curves showed a change in extraction rate only in the initial part of the curve (approximately 20–30%) and more when the flow rate increased from 0.25 to 1 mL/min. At higher flow intensities (1 and 2 mL/min), there was only a slight difference in extraction velocities indicating that initially the supercritical CO_2 extraction process was controlled by desorption kinetics for both clove essential oil compounds. However, the desorption kinetic model gave a good fit for all the intensities of the tested CO_2 currents.

In fact, they found that the extraction of essential oil from cloves during SFE is controlled by transferring the internal mass of the dissolved analyte into the solid matrix, but in SWE it is controlled by transferring the external mass of the analyte to the solvent. This is partly explained by the fact that SWE is more inclined to extract organic matter than SFE. They hypothesized that SWE was more effective than supercritical CO_2 in changing sample matrix and moving analytes from their original graft sites.

Ankpankol et al. (2007) performed anti-cancer damnacanthal extraction from *Morinda citrifolia* roots with subcritical water [3]. They performed the extractions at different temperatures and intensities and fitted the data to determine the extraction mechanism with the four models mentioned. To quantitatively compare the extraction models, the researchers considered the mean error percentage between the experimental data and the models (Table 4.1). Based on the results in Table 4.1, the K_D model was suitable for describing extraction at all flow intensities. On the other hand, unicellular and bi-spatial desorption models described the extraction data well at high-volume flow intensities. Of all the methods, the external mass transfer model has the best description of the laboratory data. Thus, in general, a combination of dissociation coefficient (K_D) and external mass transfer gives a good description of extraction with damnacanthal subcritical water, while the kinetic desorption model describes the extraction behavior at high current intensities.

Therefore, the extraction process may be controlled by a combination of different processes, and the precise mechanisms that

TABLE 4.1
Percentage of Relative Absolute Errors between Laboratory Data and Extraction Model Results

Model	Parameter Model	% Mean Absolute Errors at Difference Flow Rate (mL/min)				
		1.6	1.6	2.4	3.2	4
Partition coefficient	K_D	2.27	2.27	2.69	3.44	2.26
One-site kinetic desorption	k	8.66	8.66	3.11	3.07	0.95
Two-site kinetic desorption	K_1, k_2	8.57	8.57	3.12	2.49	0.83
External mass transfer	K_D, $k_e a_p$	2.27	2.27	2.35	2.97	0.95

Source: Anekpankul, T., Goto, M., Sasaki, M. et al. (2007). *Separation and Purification Technology* 55(3): 343–349.

control extraction may vary depending on the extraction parameters. Khajenoori et al. (2009) modeled the extraction of thymol and carvacrol from thyme leaves with subcritical water [4] and their results suggested that the overall extraction mechanism is affected by the dissociative equilibrium of the analyte with the transfer of external mass into the liquid film.

4.2.4 Model Based on Differential Mass Balance Equations

In modeling the extraction process, more attention is paid to models based on mass balance equations for thin sections of the filled bed. Mathematically, the cylindrical extraction vessel (length L, diameter d_E) is divided into volumetric elements of finite difference with height (Figure 4.3) [5]. Mass equilibrium equations are written and arranged, and when $\Delta z \to 0$ and $\Delta t \to 0$, differential mass equilibrium equations arise. From the analysis of relevant mass transfer models and their comparison, analyte separation between solid and fluid phases, axial dispersion and mass transfer coefficients for different plant networks in filled substrates can be applied. Therefore, it will have more applications in food and engineering processes.

Ghoreshi et al. (2008) extracted and modeled mannitol from olive leaves with subcritical water [6]. Their model compares two mass transfer mechanisms: (1) Displacement flow transfer between particles and mass phase (subcritical water) with external mass transfer

Figure 4.3 Electron microscope image of anise seeds for an average particle size of 0.6 mm.

coefficient (k_f), and (2) isothermal linear equilibrium on a solid matrix surface with equilibrium coefficient (h). The equilibrium coefficient for subcritical water temperature consists of two stages of mass transfer, namely internal infiltration and adsorption/desorption equilibrium. Their simplifying assumptions include: (1) The process is one-dimensional, non-uniform and system-centric flow; (2) radial and axial scattering is negligible due to column geometry; (3) the temperature system is constant; (4) the analyte is completely dissolved in subcritical water; (5) the bed system is fixed and consists of two phases – fixed (solid) and mobile (fluid); (6) there is a linear relationship between the analyte concentration in the stationary and mobile phases; (7) fluid flow intensity, density and viscosity are constant during the extraction process; (8) the pressure and temperature gradients are negligible. They also assumed from the last two assumptions that the apparent velocity was constant along the bed. In this process, mannitol is transported by mass flow and the mass balances of the analyte in the mobile and solid phases are written as follows:

$$\varepsilon \frac{\partial c}{\partial t} = -u \frac{\partial c}{\partial z} - (k_f a)(c - c^*) \qquad (4.10)$$

$$(1 - \varepsilon) \frac{\partial c_s}{\partial t} = (k_f a)(c - c^*) \qquad (4.11)$$

With the equilibrium relationship and the following boundary conditions:

$$c_s = hc^* \qquad 0 \le z \le L,\ 0 \le t \le T \tag{4.12}$$

$$z = 0 \quad \rightarrow \quad c = 0 \tag{4.13}$$

$$t = 0 \quad \rightarrow \quad c = 0 \tag{4.14}$$

$$t = 0 \quad \rightarrow \quad c_s = c_{so} \tag{4.15}$$

a is specific solid surface area (1/cm), c is concentration of mannitol in the liquid phase (mol/cm^3 water), c_s is concentration of mannitol in the solid phase (mol/cm^3 solid), c_{so} is initial concentration of mannitol in the solid phase (mol/cm^3 solid), c^* is equilibrium concentration of mannitol in the fluid phase (mol/cm^3 water), h is equilibrium coefficient (m^3 water/m^3 olive leaves), k_f is external mass transfer coefficient (cm/s), L is bed length (m), t is time (min), u_z is the apparent velocity (cm/s), z is the axial coordinate (m) and ε is the empty bed component.

They solved these equations with two adjustment parameters, analytically and Laplace transform method for the output concentration $c(z, t)$ and solid concentration $c_s(z, t)$ and presented the Eqs. (16–26).

$$c(z,t) = \begin{cases} K_2\left[1+\exp(-K_1 t)\right], & 0 \le t \le \dfrac{\varepsilon z}{u_z} \\ K_2\left[1+e^{-K_1 t} - e^{-Az}\varphi_o\left(z, t-\dfrac{\varepsilon z}{u_z}\right) + e^{-At}\varphi_{K_1}\left(z, t-\dfrac{\varepsilon z}{u_z}\right)\right], & t \ge \dfrac{\varepsilon z}{u_z} \end{cases} \tag{4.16}$$

$$c_s(z,t) = \begin{cases} h\times\left[K_2 + K_3 e^{-K_1 t}\right], & 0 \le t \le \dfrac{\varepsilon z}{u_z} \\ h\times\left[K_2 + K_3 e^{-K_1 t} - K_2 e^{-Az}\varphi_o\left(z, t-\dfrac{\varepsilon z}{u_z}\right)\right. \\ \left. -K_3\varphi_{K_1}\left(z, t-\dfrac{\varepsilon z}{u_z}\right) + c_{so}e^{-Az}\varphi_{K_4}\left(z, t-\dfrac{\varepsilon z}{u_z}\right)\right], & t \ge \dfrac{\varepsilon z}{u_z} \end{cases} \tag{4.17}$$

$$\varphi_\eta(z,t) = e^{-\eta t} + \sqrt{\frac{\gamma\delta z}{\lambda}} \int_0^t \frac{e^{-\eta(t-s)-\delta s}}{\sqrt{s}} I_1\left(2\sqrt{\frac{\gamma\delta z s}{\lambda}}\right) ds \tag{4.18}$$

$$\lambda = u_z/\varepsilon \tag{4.19}$$

$$\gamma = k_f a/\varepsilon \tag{4.20}$$

$$\delta = (k_f a)/(h(1-\varepsilon)) \tag{4.21}$$

$$A = (k_f a)/u_z \tag{4.22}$$

$$K_1 = k_f a((1/\varepsilon) + (1/(h(1-\varepsilon)))) \tag{4.23}$$

$$K_2 = c_{so}/(h + (\varepsilon/(1-\varepsilon))) \tag{4.24}$$

$$K_3 = c_{so}/(h(1 + h(1-\varepsilon)/\varepsilon)) \tag{4.25}$$

$$K_4 = k_f a/(h(1-\varepsilon)) \tag{4.26}$$

I_1 is the modified Basel function of the first type, φ_o, φ_{K1} and φ_{K4} were defined in Eq. (4.10) with $\eta = 0$, $\eta = K_1$ and $\eta = K_4$.

Numerical integration was performed by Simpson method. The proposed mathematical model predicted well the laboratory measurements with standard deviation error (AAD) of 1.5%. The parameters of the estimated model are shown in Table 4.2. In this modeling

TABLE 4.2
Configuration Parameters and Input Data for the Extraction Model

Input Data				Configuration Parameters			
C_{so} (mol/cm³)	C_0	ε	L (cm)	U (cm/min)	$K_f a$ (min⁻¹)	h	AAD (%)
1	0	0.4	12.5	0.314	0.0098	0.55	1.2

Source: Ghoreishi, S.M. and Gholami Shahrestani, R. (2009). *Journal of Food Engineering* 93: 474–481.

method, there were two adjustment parameters ($k_f a$ (min^{-1}) and h) and the axial dispersion effect was not considered.

4.3 DESCRIPTION OF THE SELECTED MATHEMATICAL MODEL

According to the explanations provided, it was found that the use of the fourth approach, i.e., the model based on differential mass balance equations, will lead to a more comprehensive model. In fact, simultaneous solution of the equations governing the process in both solid and fluid phases at subcritical temperatures will be required. Using this model, it is possible to calculate the changes in the soluble concentration, both in the particles and in the fluid phase, at any time and place of the substrate, and to obtain the amount of product extraction. First, we briefly review the extraction model under study. This model was first developed by Goto et al. (1993) to predict analyte concentrations in both solid and solid phases [7]. Further modifications were made to this model by Dunford, Goto and Tamley (1998) [8]. Since then, this model has been used in supercritical fluid extraction. Attempts have been made to use this model for SWE of essential oils [9, 10].

It was used scanning electron microscopy (SEM) images to study the shape of plant particles. Figure 4.4 shows an electron microscope image of anise grains for an average particle size of 0.6 mm. Using measurement software, we determined the particle size of 30 particles and then their average dimension was determined. This number was very close to 0.6 mm. Microscopic shapes, on the other hand, confirmed that the particles were almost spherical. In this work, a two-phase model including solid and subcritical fluid phases was considered. The extraction vessel is considered to be cylindrical, filled with spherical solid particles of the same size (Figure 4.3).

In the SWE, the process was assumed to exist in three successive stages. First, analytes penetrate from the core of plant material to the surface. They are then extracted from the surface into the fluid and transferred to the fluid bulk. Schematic representation of these three stages and a clearer view of the bed can be seen in Figure 4.4 and Figure 4.5. Extraction speed is limited by the speed of the slowest step. In order to develop an applied mathematical model of essential oil transfer in SWE of thyme particles, some important simplifying hypotheses on which the model of process simulation was based are:

- The process of temperature and pressure is constant.
- The physical properties of subcritical water assumed constant.

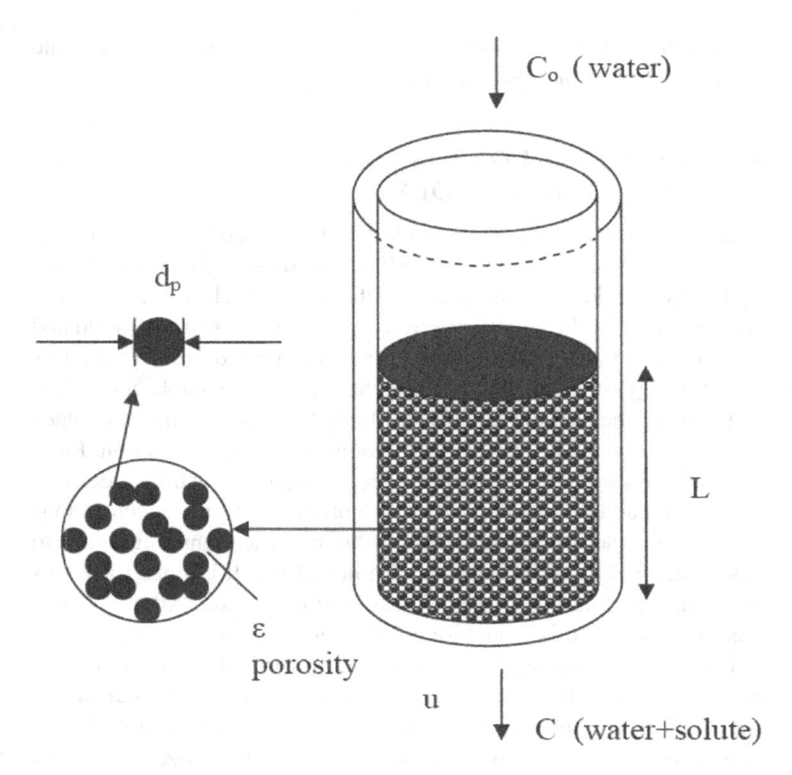

Figure 4.4 General shape of fixed bed of SWE.

(Source: Haghighi Asl, A. and Khajenoori, M. (2013). *Chapter Book, InTech*, 459–487).

- The concentration profile in the radial direction in the extraction vessel was omitted.
- Flow regime was assumed to be axial scattering.
- Essential oil was taken as a single compound and other possible effects of components in plant particles on the extraction process were ignored at the desired temperature and pressure.
- The soluble diffusion coefficient in the solid phase is assumed to be constant.
- The soluble concentration in spherical solid particles was considered independent of the θ and φ components.

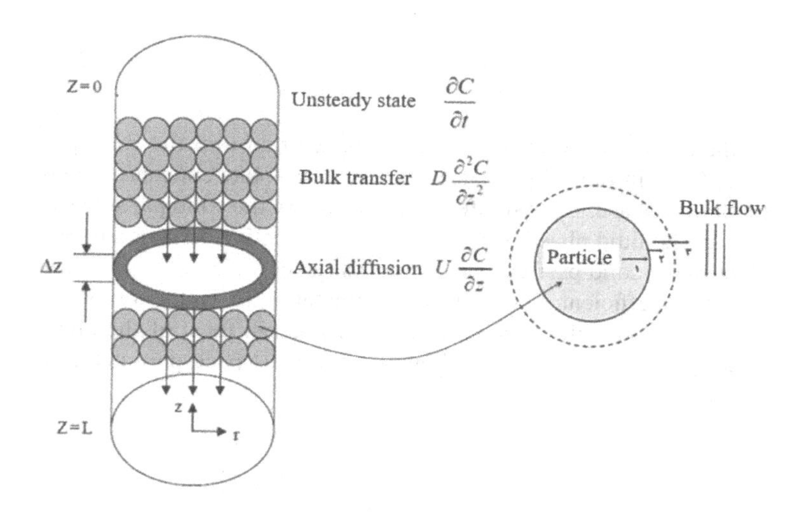

Figure 4.5 General scheme of mass balance in fixed bed and simplification of SWE: (1) Diffusion of analytes into the particle; (2) wash the analytes with a flowing fluid; (3) diffusion of analytes into the residing fluid layers outside the particles.

(Source: Khajenoori, M.; Haghighi Asl, A. and Eikani, M.H. (2014). *Ph.D. Thesis*).

According to the above hypotheses, the mass balance was in the solid phase and the subcritical water phase, both of which led to partial differential equations.

From the unstable mass balance of the solute with a constant diffusion coefficient in a spherical particle, the concentration changes with respect to the radius of the sphere at any desired time were obtained as follows:

$$\frac{\partial C_s}{\partial t} = D_m \frac{1}{r^2} \frac{\partial}{\partial r}\left(r^2 \frac{\partial C_s}{\partial r} \right) \tag{4.27}$$

$$t=0, \qquad 0 \le r \le R, \qquad C_s = C_{so} \tag{4.27-a}$$

$$t>0, \qquad r=0, \qquad \frac{\partial C_s}{\partial r} = 0 \tag{4.27-b}$$

$$t > 0, \qquad r = R, \qquad -D_m \frac{\partial C_s}{\partial r} = k_f(c_{fs} - c_f) \tag{4.27-c}$$

In the above equation, C_s is solute concentration in the solid particle, C_{so} is initial solute concentration, C_f is solute concentration in the liquid phase, C_{fs} is solute concentration in the solid particle surface in the fluid phase, R is radius, D_m is effective diffusion coefficient in the solid particle, ε is the porosity coefficient, k_f is the mass transfer coefficient, t is the time parameter and r is the coordinate of space within the sphere. The dimensionless form of the above equation in terms of dimensionless time and space parameters is as follows:

$$\frac{\partial C_s}{\partial \tau} = \frac{2}{Pe_p} \frac{L}{R} \left[\frac{1}{X^2} \frac{\partial}{\partial X} \left(X^2 \frac{\partial C_s}{\partial X} \right) \right] \tag{4.28}$$

$$\tau = 0, \qquad 0 \leq X \leq 1, \qquad C_s = C_{so} \tag{4.28-a}$$

$$\tau > 0, \qquad X = 0, \qquad \frac{\partial C_s}{\partial X} = 0 \tag{4.28-b}$$

$$\tau > 0, \qquad X = 1, \qquad -\frac{\partial C_s}{\partial X} = Bi(C_{fs} - C_f) \tag{4.28-c}$$

Pe_p is the Péclet number of solid particle $\left(\frac{2u_oR}{D_m\varepsilon}\right)$, u_o is the apparent velocity of the fluid, Bi is the Biot number $\left(\frac{k_fR}{D_m}\right)$, τ is dimensionless time $\left(\frac{u_ot}{L\varepsilon}\right)$ and X is dimensionless radius $\left(\frac{r}{R}\right)$.

Also, the mass balance for the finite element in the subcritical water phase and its boundary conditions are as follows:

$$\frac{\partial C_f}{\partial t} = D_L \frac{\partial^2 C_f}{\partial y^2} - u \frac{\partial C_f}{\partial y} - \frac{3}{R} \frac{(1-\varepsilon)}{\varepsilon} k_f(C_f - C_{ps}) \tag{4.29}$$

$$t = 0, \qquad 0 \leq y \leq L, \qquad C_f = 0 \tag{4.29-a}$$

$$t>0, \quad y=0, \quad D_L \frac{\partial C_f}{\partial y} = -u\left(C_f\big|_{z=0} - C_f\big|_{z=0}\right) \quad (4.29\text{-b})$$

$$t>0, \quad y=L, \quad \frac{\partial C_f}{\partial t} = 0 \quad (4.29\text{-c})$$

D_L is the axial dispersion coefficient and L is the height of the bed.

Also, the dimensionless form of the above equation in terms of dimensionless time and space parameters is as follows:

$$\frac{\partial C_f}{\partial \tau} = \frac{1}{Pe_b}\frac{\partial^2 C_f}{\partial Y^2} - \frac{\partial C_f}{\partial Y} - \frac{6(1-\varepsilon)L}{\varepsilon R}\frac{Bi}{Pe_p}(C_f - C_{fs}) \quad (4.30)$$

$$\tau=0, \quad 0 \le Y \le 1, \quad C_f = 0 \quad (4.30\text{-a})$$

$$\tau>0, \quad Y=0^+, \quad C_f - \frac{1}{Pe_b}\frac{\partial C_f}{\partial Y} = 0 \quad (4.30\text{-b})$$

$$\tau>0, \quad Y=1, \quad \frac{\partial C_f}{\partial Y} = 0 \quad (4.30\text{-c})$$

The dimensionless longitudinal component in the bed $\left(\frac{y}{L}\right)$, Pe_b is the number of steps for the bed $\left(\frac{u_0 L}{D_L \varepsilon}\right)$. Equations (4.28) and (4.30) will be solved by the linear equilibrium relation between the concentration in the solid phase and in the subcritical water phase at the surface which is assumed as follows:

$$C_{fs} = k_p C^{+}{}_{ss} \quad (4.31)$$

C_{fs} is the concentration of analyte in the subcritical water phase at the particle level, C_{ss} is the concentration of analyte in the solid phase in equilibrium with the subcritical water phase and k_p is the equilibrium separation coefficient of the analyte between the solid and the water phase. Equations (4.28), (4.30) and (4.31) will be solved simultaneously with the three unknowns C_s, C_f and C_{fs}.

4.4 SIMULATION METHOD

Equations (4.28a–c) and (4.30a–c) are a set of partial differential equations and the solution of these equations can be obtained by numerical methods. Different numerical methods such as polynomial approximations, single perturbation methods, finite difference methods and orthogonal collocation methods are available to solve partial differential equations. Finite difference methods and orthogonal collocation are widely used to solve various chemical engineering problems [11].

Solving equations can be obtained by the finite difference method of the Crank Nicolson. Although this may not be the best choice for this problem, this method provides stable and accurate computational solutions compared to other numerical methods such as orthogonal collocation. The set of finite difference equations is solved simultaneously with the Thomas algorithm [11].

4.5 SOLUTION OF THE PARTIAL DIFFERENTIAL EQUATION

4.5.1 Essential Oil Concentration Curve in Bulk Fluid

As mentioned earlier, the dimensionless mass balance in the fluid mass phase is as follows:

$$\frac{\partial C_f}{\partial \tau}=\frac{1}{Pe_b}\frac{\partial^2 C_f}{\partial Y^2}-\frac{\partial C_f}{\partial Y}-\frac{6(1-\varepsilon)L}{\varepsilon R}\frac{Bi}{Pe_p}(C_f-C_{fs})$$

The last sentence of the above relation can be written based on Stanton:

$$\frac{L(1-\varepsilon)k_f a}{\varepsilon u}=\frac{L(1-\varepsilon)k_f a}{u_o}=St$$

Therefore, it can be written as:

$$\frac{\partial C_f}{\partial \tau}=\frac{1}{Pe_b}\frac{\partial^2 C_f}{\partial Y^2}-\frac{\partial C_f}{\partial Y}-St(C_f-C_{fs}),\ C_{fs}=C_s|_{X=1} \quad (4.32)$$

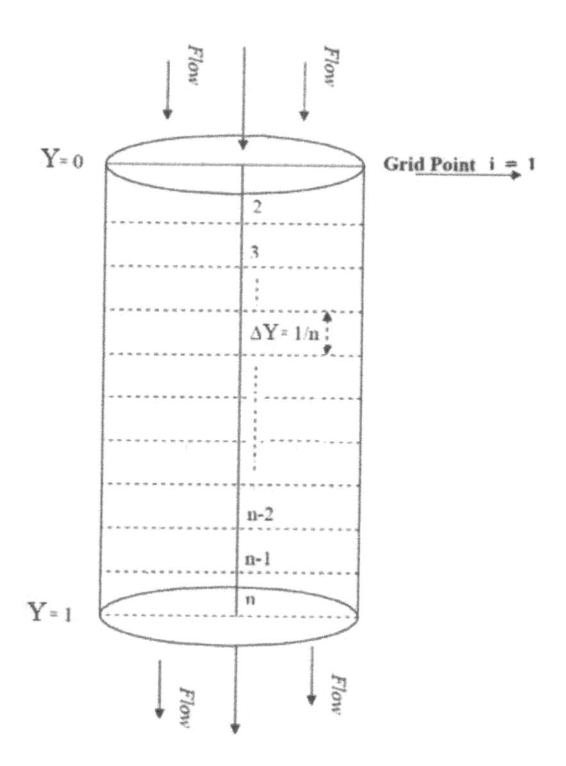

Figure 4.6 Sequencing of mesh points for finite difference method. (Source: Khajenoori, M.; Haghighi Asl, A. and Eikani, M.H. (2014). *Ph.D. Thesis*).

To solve this equation with known boundary and initial.conditions, the system was divided into n equal parts as shown in Figure 4.6 at a distance of $1 \geq Z \geq 0$ [5]. The Crank Nicholson's finite difference method was used to approximate the first- and second-degree derivatives of the C_f dimensionless concentration.

i and j are the axial and time counter, respectively.

$$\frac{\partial C_f}{\partial \tau} = \frac{(C_{f,i,j+1} - C_{f,i,j})}{\Delta \tau}$$

$$\frac{\partial^2 C_f}{\partial Y^2} = \frac{1}{2}\left[\frac{C_{f,i+1,j} - 2C_{f,i,j} + C_{f,i-1,j}}{\Delta Y^2} + \frac{C_{f,i+1,j+1} - 2C_{f,i,j+1} + C_{f,i-1,j+1}}{\Delta Y^2}\right]$$

$$\frac{\partial C_f}{\partial Y}=\frac{1}{4}\left[\frac{C_{f,i+1,j}-C_{f,i-1,j}+C_{f,i+1,j+1}-C_{f,i-1,j+1}}{\Delta Y}\right]$$

So, we have:

$$-K_1C_{f,i-1}^{j+1}+L_1C_{f,i}^{j+1}+M_1C_{f,i+1}^{j+1}=-K_1C_{f,i-1}^{j}+N_1C_{f,i}^{j}$$
$$-M_1C_{f,i+1}^{j}+StC_s\big|_{X=1}$$
$$1\le i\le n \tag{4.33}$$

That:

$$K_1=\frac{1}{2\,Pe(\Delta Y)^2}+\frac{1}{4\,\Delta Y}$$

$$L_1=\frac{1}{\Delta\tau}+\frac{1}{2\,Pe(\Delta Y)^2}+St$$

$$M_1=-\frac{1}{2\,Pe(\Delta Y)^2}+\frac{1}{4\,\Delta Y}$$

$$N_1=\frac{1}{\Delta\tau}-\frac{1}{Pe(\Delta Y)^2}$$

And:

$$j=0\quad 0\le i\le n+1\quad Cf^0{}_i=0 \tag{4.33-a}$$

$$C_{f,0}^{j+1}=\frac{C_{f,1}^{j+1}}{1+(\Delta Y)\,Pe} \tag{4.33-b}$$

$$C_{f,n+1}^{j+1}=C_{f,n}^{j+1} \tag{4.33-c}$$

$C_s\big|_{X=1}$ is the concentration of essential oil on the surface of a particle that is obtained by solving the equilibrium equation of the particle mass. Equation (4.33) is solved using Thomas algorithm to obtain a two-dimensional matrix C_f.

4.5.2 Essential Oil Concentration Curve in Solid

As mentioned before, the dimensionless mass balance in the solid phase is as follows:

$$\frac{\partial C_s}{\partial \tau} = \beta\left(\frac{\partial^2 C_s}{\partial X^2} + \frac{2}{X}\frac{\partial C_s}{\partial X}\right), \quad \beta = \frac{2}{Pe_p}\frac{L}{R}$$

And with Crank Nicholson's finite difference method relations for approximating first- and second-order derivatives of dimensionless C_s, we have:

$$-K_2 Cs_{i-1}^{j+1} + L_2 Cs_i^{j+1} + M_2 Cs_{i+1}^{i+1} = K_2 Cs_{i-1}^{j} + N_2 Cs_i^{j} - M_2 Cs_{i+1}^{i+1} \tag{4.34}$$

$$K_2 = \frac{-\beta}{2(\Delta y)^2} + \frac{1}{2y\Delta y}$$

$$L_2 = \frac{1}{\Delta t} + \frac{\beta}{(\Delta y)^2}$$

$$M_2 = \frac{-\beta}{2(\Delta y)^2} - \frac{\beta}{2y\Delta y}$$

$$N_2 = \frac{1}{\Delta t} - \frac{\beta}{(\Delta y)^2}$$

And:

$$Cs_1^{j+1} = Cs_0^{j+1} \tag{4.34-a}$$

$$Cs_n^{j+1} = Cs_{n+1}^{j+1} \tag{4.34-b}$$

$$Cs_n^{j+1} = \frac{sh \cdot y \cdot C_f + Cs_n^{j+1}}{1 + sh \cdot \Delta y} \tag{4.34-c}$$

This model will be able to show the effect of various process parameters such as temperature, particle size and flow rate on extraction efficiency. Also, concentration curves in solid phase and subcritical phase are extracted in each part of the bed and are available at any time.

As mentioned, to solve the equations with known boundary and initial conditions, the system was divided into *n* equal parts at a distance of $0 \geq Z \geq 1$ in the form of finite difference (Figure 4.6). At each step (*n*), it is assumed that all particles have the same extraction conditions and the composition of bulk fluid (free flow, C_f) is the same.

In fact, the dispersed plug-flow extractor is replaced by a series of stirred extractors. By placing Eq. (4.31) in Eqs. (4.29) and (4.30), the solution of algorithm is followed as follows:

Step 1: The profile of C_f concentration (t, z) in the fluid phase was considered. Therefore, the value of C_f in Eq. (4.29) in each part of the bed was obtained from this profile and was constant.

Step 2: Using Eq. (4.29), the concentration profile in the solid particle was obtained.

Step 3: Having the solute concentration at the particle surface, Eq. (4.30) was solved. The solution was a profile of a new concentration in the fluid phase.

Step 4: The calculated concentration profile was compared with the assumed concentration profile.

If the difference between the two concentration profiles was less than the convergence criterion, this method was approved for the next time step, and if not, the assumed concentration profile was replaced with the calculated value and the loop was continued. In both solid phase and subcritical phases, three-diagonal matrices are obtained which can be easily solved by Thomas algorithm at any time stage [11].

4.6 ESTIMATION OF MODEL PARAMETERS AND PHYSICAL PROPERTIES

Model parameters include mass transfer coefficient of bulk phase, axial dispersion coefficient in subcritical phase and the effective diffusion coefficient in the solid is that the last parameter was selected as the setting parameter.

4.6.1 Estimation of the Equilibrium Dissociation Coefficient of the Analyte

The equilibrium separation coefficient can be estimated experimentally. The extraction system is exactly the same as our previous laboratory work [12]. The extract cell was filled with 10 g of dried thyme leaves. Equilibrium experiments were performed with 20, 40, 60, 80 and 100 mL of water (different ratio of particle/solvent). Extractions at temperature 150°C and 20 bar pressure were performed. The flow rate inside the system should be low enough that the composition of the cell output is very close to the equilibrium state of the solid analyte and the liquid solution [13]. Based on the equilibrium tests in this work, the maximum current intensity in these experiments cannot be more than 0.5 mL/min. In these extractions, the concentration of thymol was determined in two phases.

To determine the amount of thymol in the solvent, the extract sample (20 mL) was completely evaporated at 60°C. The residual mass was dissolved in 20 mL of hexane in a cold bath. The solution was filtered on a 0.25 μm membrane and stored in a glass in a refrigerator until injection into GC. However, to determine the amount of thymol in the leaves, the extracted leaves were dried in a 120°C oven for 3 h. Dried leaves were crushed and meshed with standard sieves (0.5 mm mesh). Ten grams of this sample was taken with 50 mL of distilled water and boiled with a magnetic stirrer for 30 min. The resulting mixture was filtered through filter paper. Five milliliter of the filtered sample was stored in the sample container in the refrigerator until injection into GC.

The equilibrium thymol concentration between the phases is shown in Figure 4.7 [5]. As a result, the separation constant between the two phases is estimated to be $kp = 0.96$.

4.6.2 Estimation of Mass Transfer Coefficient of Bulk Phase

The mass transfer coefficient was estimated as follows using the experimental relationship reported by Vakav [13]:

$$Sh = 2 + 1.1\,Sc^{\frac{1}{3}}Re^{0.6} \tag{4.35}$$

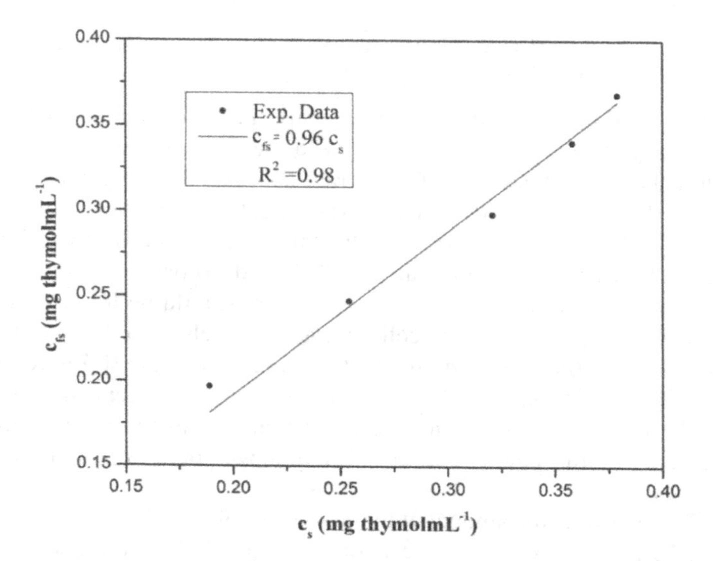

Figure 4.7 Equilibrium thymol concentration obtained in different particle/solvent ratio.

(Source: Khajenoori, M.; Haghighi Asl, A. and Eikani, M.H. (2014). *Ph.D. Thesis*).

Sh is Sherwood number $\left(\frac{d_p k_f}{D_f}\right)$, *Re* is Reynolds number $\left(\frac{u\rho_f d_p}{\mu_f}\right)$, *Sc* is Schmidt number $\left(\frac{\mu_f}{\rho_f D_f}\right)$, D_f is diffusion coefficient of soluble organic matter at subcritical water (cm²/s) (cm²/s), u is apparent fluid velocity (cm/s), ρ_f and μ_f is density (g/cm³) and viscosity of the fluid (g/cm³ . s), respectively. So, we have:

$$k_f = \frac{ShD_f}{dp}$$

4.6.3 Estimation of Solute Diffusion Coefficient in Fluid

To estimate this value, the Wilk-Chang relationship can be used [14]:

$$D^0 f = \frac{7.4 \times 10^{-8}}{\mu V_A^{\ 0.6}} (\varphi M_B)^{0.5} T \tag{4.36}$$

$D^0{}_f$ is solute diffusion coefficient at very low concentrations in water (cm²/s), M_B is molecular weight of fluid (g/mol), T is temperature (K), φ is cohesion coefficient for fluid (2.26 for water and 1.5 for ethanol) and V_A are the molar volume of solute at normal boiling point (cm³/mol). V_A can be estimated using the Thin Callus relationship [15]:

$$V_A = 0.285(V_c)^{1.048} \tag{4.37}$$

V_c is the critical volume (cm³/mol) that is estimated using the Jubek and Reed (1987) relationship [16]. Very few studies have been performed on obtaining data on changes in mass transfer properties in the subcritical fluid region. However, the relationships of the diffusion coefficients in the liquid must be extrapolated or developed to be applicable to the subcritical fluid region. Suitable studies have been presented to predict diffusion coefficient in subcritical water, but there is very little information on the comparison of calculated values. New experimental data on the finite temperature range for carbohydrates and amino acid monomers suggest the use of the Thien equation [17]:

$$\frac{D^0{}_f\,(T_2)}{D^0{}_f\,(T_1)} = \left(\frac{T_c - T_1}{T_c - T_2}\right)^n \tag{4.38}$$

The critical temperature of solvent T_c and n for water is 6.

Estimation of axial dispersion coefficient in subcritical phase was estimated using the following equation [18]:

$$Pe_{pd} = 1.634 Re^{0.265} Sc^{-0.919} \tag{4.39}$$

That:

$$D_L = \frac{udp}{Pe_{pd}\varepsilon} \tag{4.40}$$

4.6.4 Physical Properties

Bulk density (ρ_b) of particles is defined as the ratio of mass to total volume of the sample, which is modified using the standard test and weight method by filling a 100 mL container with a sample (has the same average size) by a same rate. Then the container is weighed. No manual separation of the sample should be performed. Bulk density is obtained from the sample mass and the volume of the container [18].

True density (ρ_t) of particles is defined as the ratio of the mass of the sample to the volume occupied by the sample, which can be determined using an electronic scale (0.001 g) and a pycnometer (50 + 0.1 mL) and the method of liquid placement. [19]. Xylene (0.862 ± 0.001 g/cm^3) is used due to the water absorption by the seeds and reduction of volume. Due to the short duration of the experiment, xylene uptake is assumed to be negligible [20].

Bed porosity (ε) is the part of the empty space of the seed mass that is obtained using the following equation [21]:

$$\varepsilon = \left(\frac{\rho_t - \rho_b}{\rho_t}\right) \times 100$$

The volume of one seed (V) (mm) is obtained using the following equation [22]:

$$V = \left(\frac{m}{\rho_t}\right) \times 10^3$$

Mass of a seed (m, g) of seeds are used to determine the true density. The equivalent diameter (d_p) of a sphere that are the same volume as the seeds is obtained as follows:

$$d_p = \left(\frac{6V}{\pi}\right)^{1/3}$$

Therefore, the mass transfer surface (a) of a seed is equal to:

$$a = \pi d_p^2$$

Bulk density: The density of water, at high temperature and pressure from 273 to 573 K, was assumed the density of saturated water. Using the data of water saturation tables and with the best fit of the available data, the following relation with 0.06% relative error is suggested (Figure 4.8) [5]:

$$\rho = -0.002733 \times T^2 + 1.353 \times T + 835.4 \qquad (4.41)$$

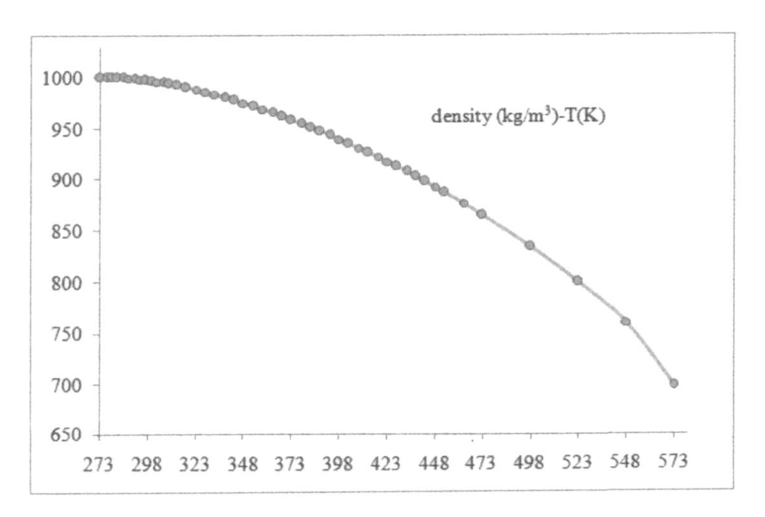

Figure 4.8 Water density (kg/m³) in terms of temperature (K). (Source: Khajenoori, M.; Haghighi Asl, A. and Eikani, M.H. (2014). *Ph.D. Thesis*).

Bulk viscosity: The viscosity of water at 273 to 573 K was estimated from the available data using the following relation with 0.01% relative error (Figure 4.9) [16]:

$$\mu = 6.41\times10^{4}\ \exp\left(-0.03964\times T\right)+2.789 \times \exp\left(-0.006304\times T\right) \tag{4.42}$$

Where μ is the viscosity ($Pa \cdot s$) and T is the temperature (K). It was assumed that the viscosity and density of water are independent of pressure.

In this case, since the extraction curve is defined as a dimensionless variable, the cumulative amount of extracted essential oil is the total amount of essential oil in thyme leaves, so it is necessary to integrate the output current intensity as a function of time as follows:

$$\text{Extraction yield} = \frac{m_{\text{comp}}}{m_{\text{comp}\infty}} = \frac{\int_0^t c\ dt}{\int_0^\infty c\ dt} \tag{4.43}$$

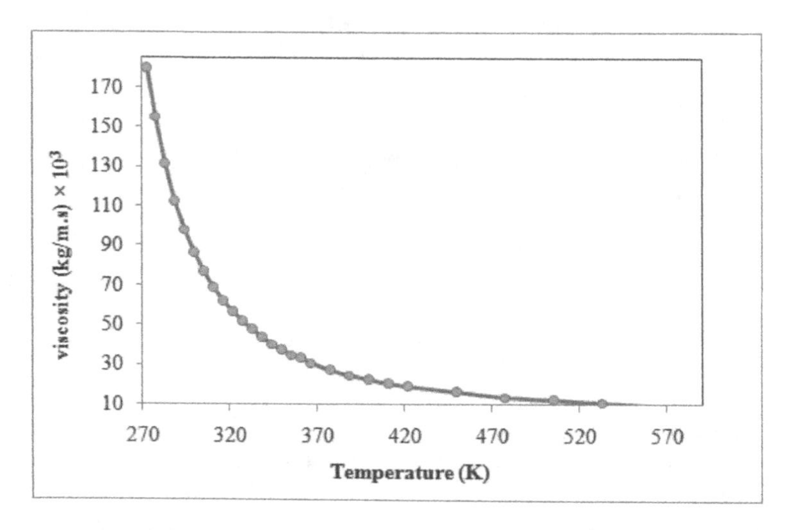

Figure 4.9 Water viscosity (kg/m^3) in terms of temperature (K). (Source: Khajenoori, M.; Haghighi Asl, A. and Eikani, M.H. (2014). *Ph.D. Thesis*).

m_{comp} is the weight of the extracted compound (g) and $m_{comp\infty}$ is the maximum weight of the extracted compound (g).

REFERENCES

1. Haghighi Asl, A., Khajenoori, M. (2013). Subcritical water extraction, *Mass transfer-advances in sustainable energy and environment oriented numerical modeling, InTech*, Rijeka.
2. Kubatova, A., Jansen, B., Vaudoisot, J.F., Hawthorne, S.B. (2002). Thermodynamic and kinetic models for the extraction of essential oil from savory and polycyclic aromatic hydrocarbons from soil with hot (subcritical) water and supercritical CO_2, *Journal of Chromatography A.*, 975 (1): 175–188.
3. Anekpankul, T., Goto, M., Sasaki, M., Pavasant, P., Shotipruk, A. (2007). Extraction of anti-cancer damnacanthal from roots of *Morinda citrifolia* by subcritical water. *Separation and Purification Technology*, 55(3): 343–349.
4. Khajenoori, M., Haghighi Asl, A., and Hormozi, F. (2009). Proposed models for subcritical water extraction of essential oils, *Chinese Journal of Chemical Engineering*, 17 (3): 359–365.

5. Khajenoori, M., Haghighi Asl, A. and Eikani, M.H. 2014. *Investigation of thermosynthetic parameters of plant extraction with subcritical water. Ph.D. Thesis*, Semnan University, Iran.
6. Ghoreishi, S.M., Gholami Shahrestani, R. (2009). Subcritical water extraction of mannitol from olive leaves, *Journal of Food Engineering*, 93: 474–481.
7. Goto, M., Sato, M., Hiroshe, T. (1993) Extraction of peppermint oil by supercritical carbon dioxide. *Journal of Chemical Engineering of Japan*, 26: 401–407.
8. Dunford, N. T., Goto, M., Temelli, F. (1998) Modeling of oil extraction with supercritical CO_2 from Atlantic mackerel (*Scomber scombrus*) at different moisture contents. *The Journal of Supercritical Fluids*, 13: 303–309.
9. Khajenoori, M., Omidbakhsh, E., Hormozi, F., Haghighi Asl, A. (2009). CFD modeling of subcritical water extraction, the 6th International chemical Engineering Congress (IChEC), Kish Island, Iran.
10. Khajenoori, M., Haghighi Asl, A. and Eikani, M.H. (2010). Modeling of superheated water extraction of essential oils. The 13th Iranian National Chemical Engineering Congress & 1st International Regional Chemical and Petroleum Engineering Kermanshah, Iran.
11. Rice, R. G., Do, D. D. (1995). *Applied mathematics and modeling for chemical engineers.* Chapter 3, John Wiley & Sons, Inc. 576–586.
12. Wakao, N., Kaguei, S. (1982). *Heat and mass transfer in packed beds*, 1st ed. Routledge, London, UK, 192–205.
13. Wilke, C.R., Chang, P. (1955). Correlation of diffusion coefficients in dilute solutions, *AIChE Journal* 1, 264–270.
14. Bocquet, S., Romero, J., Sanchez, J., Rios, G.M. (2007). Membrane contactors for the extraction process with subcritical carbon dioxide or propane: Simulation of the influence of operating parameters. *Journal of Supercritical Fluids*, 41: 246–256.
15. Joback, K. G., and Reid, R. C. (1987). Estimation of pure component properties from group contributions. *Chemical Engineering Communication*, 57: 233–243.
16. Perry, R., Green, D.W., Maloney, J.O. (1984). *Perry Chemical Engineer's Handbook*, 5th ed. McGraw-Hill, New York.
17. Tan, C. S. and Liou, D. C. (1989). Axial dispersion of Supercritical carbon dioxide in packed beds. *Industrial & Engineering Chemistry Research*, 28: 1246–1250.
18. Eikani, M.H. and Rowshanzamir, S. (2004). Modeling and simulation of superheated water extraction of essential oils, CHISA, 16th International Congress of Chemical and Process Engineering, 4563–4567.

19. Singh, K.K., Goswami, T.K. (1996). Physical properties of Cumin seed. *Journal of Agricultural Engineering Research*, 64: 93–98.
20. Mohsenin, N.N. (1970). *Physical Properties of Plant and Animal Materials*. Gordon Breach Science Publishers, New York.
21. Giner, S.A., Calvelo, A. (1987). Modelling of wheat drying in fluidized beds. *Journal of Food Science*, 52: 1358–1363.
22. Ixtaina, V.Y., Nolasco, S.M., Tomas, M.C. (2008). Physical properties of chia (*Salvia hispanica L.*) seeds, *Industrial Crops and Products*, 28: 286–293.

SUBJECT INDEX

A

Accelerated solvent extraction (ASE), 3
Additives, 54
Antioxidants, 5, 9, 22, 45
Anthocyanins, 20, 46
Apparent velocity, 53, 94, 95
Aromatic, 1, 5, 7, 8, 21, 27, 37, 40, 45, 46, 64, 70
Axial dispersion coefficient, 101, 106, 109

B

Bed porosity, 110
Biot number, 100
Bulk density, 109, 110
Bulk viscosity, 111

C

Cavitation bubbles, 11, 14, 15
Clevenger, 3
Controlled pressure drop process, 3, 17
CPA EOS, 70, 71
Crank Nicolson, 102

D

Dean–Stark method, 3
Density, 21, 22, 24, 29, 38, 46, 88, 94, 108, 109, 110
DIC process, 19, 20, 21
Dielectric constant model, 69
Diffusion coefficient, 88, 98, 99, 100, 106, 108, 109
Dynamic, 3, 23, 52, 53, 54, 62, 64, 65, 66

E

Essential oils, 1, 2, 4, 5, 6, 7, 8, 20, 21, 24, 25, 26, 28, 37, 41, 44, 45, 46, 47, 90, 97
Equations of State (EOS), 70

F

Finite difference, 93, 102, 103, 105, 106
First approximation, 68
Flame atomic-absorption spectrometer (FAAS), 14
Flow Rate, 23, 25, 27, 53, 54, 63, 64, 65, 66, 88, 92, 106, 107
Food ingredients, 47

G

Gas chromatography-mass spectrometry (GC-MS), 14
Green solvents, 37
Group-contribution, 74

H

Heterogeneous chemical groups, 1, 2
Hydrophobic organic compounds, 53, 77

K

Kjeldahl method, 14

M

Microwave-assisted solvent extraction (MASE), 5
Microwave extraction, 3, 5, 7, 8, 9, 10, 15, 16
Model based on differential mass balance equations, 87, 97
Monoterpene, 1, 30

N

Non-polar analytes, 8, 15
Nutritional compounds, 45

P

Peclet number, 100
Phenolic compounds, 17, 34, 45
Polychlorinated biphenyls (PCBs), 45
Polycyclic aromatic hydrocarbons (PAHs), 40, 45

R

Reynolds number, 53, 108

S

Schmidt number, 108
Sedimentary environmental samples, 45
Semi-experimental models, 67
Sesquiterpene hydrocarbons, 1
Sherwood number, 53, 108
Single-location kinetic models, 87
Sediments, 3, 25, 39, 45
Soil, 40, 45, 90
Solubility of Subcritical Water, 51
Solute fugacity, 72
Soxhlet, 2, 3, 9, 14, 15, 16, 17, 24, 27, 28, 30, 42, 43, 44, 47
Solvent-free microwave extraction (SFME), 5
Solvent type, 41, 52
Static Mode, 53
Subcritical water, 3, 5, 25, 26, 27, 28, 29, 30, 31, 37, 38, 39, 40, 41, 47, 51, 62, 70, 85, 90, 91, 92, 93, 94, 97, 99, 100, 101, 108, 109
Subcritical water extraction (SWE), 3, 25, 37
Supercritical fluid extraction (SFE), 3, 42, 43
Superheated water, 37, 41
Synthetic models, 87

T

Thermodynamic Framework, 71
Thermodynamic model, 87
Thermodynamic separation model with external mass transfer resistance, 87
Thomas algorithm, 102, 105, 106
True density, 110
Two-dimensional kinetic models, 87

U

Ultrasound-assisted extraction (UAE), 3, 10
Ultrasonic extraction, 9, 15, 16, 17, 36
UNIFAC-Based Models, 74, 77
UNIFAC model, 74, 75, 77, 78,

Z

Zeroth approximation, 67